A guide to advanced manufacturing in electronics

30125 00437207 3

Books are to be returned on or before
the last date below.

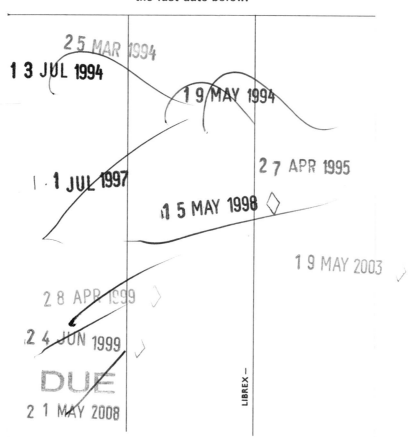

A guide to advanced manufacturing in electronics

P H du Feu

Inbucon Technology, The Venture Centre,
Sir William Lyons Road, Warwick Science Park,
Coventry CV4 7EZ

Produced as part of a joint study project by the UK
electronics industry and the Department of Trade & Industry

Published by: Peter Peregrinus Ltd., London, United Kingdom

© Crown copyright, 1988

All rights reserved. No part of this publication may be reproduced, stored in a retrieval system or transmitted in any form or by any means — electronic, mechanical, photocopying, recording or otherwise — without the prior permission of the publisher.

While the author and the publishers believe that the information and guidance given in this work are correct, all parties must rely upon their own skill and judgment when making use of them. Neither the author nor the publishers assume any liability to anyone for any loss or damage caused by any error or omission in the work, whether such error or omission is the result of negligence or any other cause. Any and all such liability is disclaimed.

The cover photograph shows ICL's most advanced printed circuit board using surface mount technology

ISBN 0 86341 131 2

Printed in England by Black Bear Press Ltd., Cambridge.

Contents

	Page
Preface	ix
Foreword	xi

1 Advanced manufacturing in electronics—AMIE — 1
  1.1 Introduction — 1
  1.2 What is AMIE? — 3
    1.2.1 Technology — 3
    1.2.2 Components — 4
    1.2.3 Related approaches — 4
  1.3 Summary — 6
  1.4 Bibliography — 6

2 Justification — 7
  2.1 Introduction — 7
  2.2 Benefits of AMIE — 7
    2.2.1 Higher quality and reliability of product — 8
    2.2.2 Higher quality and reliability of service — 8
    2.2.3 Reduced tendering time — 8
    2.2.4 Shorter design market-availability time — 8
    2.2.5 Decreased production cycle time — 9
    2.2.6 Direct cost saving — 9
    2.2.7 Improved company image — 10
    2.2.8 New market opportunities — 10
    2.2.9 The only way — 10
    2.2.10 Summary of benefits — 10
  2.3 Justification case studies — 11
  2.4 The cost of AMIE — 11
  2.5 Financial evaluation methods — 12
    2.5.1 Payback methods — 12
    2.5.2 Discounting techniques — 12
    2.5.3 Comparison of methods — 14
    2.5.4 Sensitivity analysis — 14
    2.5.5 Risk analysis — 14
  2.6 Tools for financial evaluation — 14
  2.7 Summary — 15
  2.8 Bibliography — 16

3 Design — 17
  3.1 Introduction — 17
  3.2 Design process — 17
  3.3 Circuit capture — 19
  3.4 Simulation — 21
    3.4.1 Analogue simulation — 21
    3.4.2 Logic simulation — 21

				Page
		3.4.3	Fault simulation	22
		3.4.4	Timing	22
		3.4.5	Thermal	23
	3.5	Layout		23
		3.5.1	Component placement	23
		3.5.2	Track routing	24
	3.6	Documentation		25
	3.7	Post-processing		25
	3.8	Libraries and data bases		25
	3.9	Hardware		26
	3.10	Plotting		27
	3.11	Summary		27
	3.12	Bibliography		27

4 Manufacture   28
    4.1 Introduction   28
    4.2 Assembly methods   29
        4.2.1 Manual assembly   29
        4.2.2 Semi-automatic   29
        4.2.3 Flow line   30
        4.2.4 Medium volume/batch equipment   30
        4.2.5 Dedicated equipment   30
        4.2.6 Surface-mount device placement   30
        4.2.7 Robotics   31
        4.2.8 Equipment selection   31
    4.3 Solder and cleaning   31
        4.3.1 Flow soldering   32
        4.3.2 Re-flow soldering   32
        4.3.3 PCB cleaning   32
        4.3.4 Solder masking   33
        4.3.5 Conformal coating   33
    4.4 Storage and Handling   33
        4.4.1 Storage   33
        4.4.2 Handling   34
    4.5 Production environment   34
    4.6 Summary   34
    4.7 Bibliography   34

5 Test   35
    5.1 Introduction   35
    5.2 Test and inspection tools   35
        5.2.1 Goods-inwards inspection procedures   35
        5.2.2 Machine vision   36
        5.2.3 In-current test   37
        5.2.4 Functional test for consistency   38
        5.2.5 Combinational test for consistency   38
        5.2.6 Environmental stress screening   38
    5.3 Re-work   39
    5.4 Test area data bases   39
    5.5 Strategy   40
    5.6 Summary   41
    5.7 Bibliography   41

		Page
6	**Manufacturing resource planning**	42
	6.1 Introduction	42
	6.2 Business problems	43
	6.3 MRPII: Practical benefits	44
	6.4 MRPII: Functional description	44
	6.4.1 Bill of materials	44
	6.4.2 Material requirements planning	44
	6.4.3 Master schedule	44
	6.4.4 Shop-floor control	44
	6.4.5 Customer order entry	45
	6.4.6 Purchasing	45
	6.4.7 Stock control	45
	6.4.8 Job costing	45
	6.5 Systems integration	45
	6.6 MRPII: Summary	45
	6.7 Bibliography	46
7	**Just-in-time manufacture**	47
	7.1 Introduction	47
	7.2 Implementation	48
	7.2.1 Top management	48
	7.2.2 People development	48
	7.2.3 Total quality	48
	7.2.4 Waste elimination	49
	7.2.5 Manufacturing route simplification	49
	7.2.6 Visibility of problems	52
	7.2.7 Continuous improvement	52
	7.2.8 Supplier relationships	52
	7.2.9 Summary	53
	7.3 Computer systems	53
	7.4 Summary	55
	7.5 Bibliography	55
8	**Support tools**	56
	8.1 Introduction	56
	8.2 Interconnection, interfacing and integration	56
	8.2.1 Interconnection	57
	8.2.2 Interfacing	57
	8.2.3 Integration	57
	8.3 Communications and networking standards	57
	8.4 Data bases and data management	59
	8.5 Data standards	60
	8.5.1 EDIF: Electronic Design Interchange Format	61
	8.5.2 IGES: Initial Graphics Exchange Specification	61
	8.6 Bar codes	61
	8.7 Fourth-generation languages	62
	8.8 Bibliography	63
9	**Implementation**	64
	9.1 Introduction	64
	9.2 Initiating action	64
	9.3 Business goal definition	66
	9.4 Strategy development	66

			Page
	9.5	Modelling and simulation	68
		9.5.1 Function modelling	69
		9.5.2 Information modelling	69
		9.5.3 Dynamic modelling	70
		9.5.4 Selection of planning tools	70
	9.6	Evaluation	71
	9.7	Preparation and installation	74
	9.8	Auditing	76
	9.9	Summary	76
	9.10	Bibliography	76
10	**Typical installations**		**78**
	10.1	Introduction	78
	10.2	Case study 1	78
		10.2.1 Background	78
		10.2.2 Design and process development group	79
		10.2.3 Manufacturing plants	79
		10.2.4 Summary	80
	10.3	Case study 2	81
		10.3.1 Background	81
		10.3.2 Manufacturing research centre	81
		10.3.3 Small-batch manufacturing plant	82
		10.3.4 Volume manufacturing plants	82
	10.4	Summary	83
11	**Component technology**		**85**
	11.1	Introduction	85
	11.2	Surface mount	85
		11.2.1 Surface mount: The technology	85
		11.2.2 Surface mount: The benefits	86
		11.2.3 Surface mount: Market trends	87
		11.2.4 Surface mount: Impact on design	87
		11.2.5 Surface mount: Substrate requirements	87
		11.2.6 Surface mount: Component sourcing	87
		11.2.7 Surface mount: Production	87
		11.2.8 Surface mount: Test strategy	87
	11.3	Application specific integrated circuits	88
		11.3.1 ASICs: Introduction	88
		11.3.2 ASICs: Device description	90
		11.3.3 Programmable logic devices	90
		11.3.4 Gate array	91
		11.3.5 Standard cell and full custom	92
	11.4	Summary	93
	11.5	Bibliography	93
12	**A look ahead**		**94**
	12.1	Introduction	94
	12.2	Competitive pressure	94
	12.3	Technology of the future	95
		12.3.1 Design	95
		12.3.2 Manufacture	95
		12.3.3 Computers and integration	95
		12.3.4 Component and packaging technology	96
	12.4	Summary	96
	12.5	Bibliography	96
13	**Glossary of terms**		**97**

Preface

This book is one of the results of a joint initiative of the Department of Trade and Industry and the Electronic Engineering Association with support from other Trade Associations.

Some of the major UK Electronic Companies have contributed their time, knowledge and experience to produce a unique book which covers a wide scope of subjects from technology to implementation strategy for achieving successful exploitation of Advanced Manufacturing in Electronics.

The aim of the book is to increase awareness of what is possible with the new technologies and also to encourage much wider use through an understanding of the benefits and how to avoid the potential pitfalls.

AMIE and the movement towards integration is essential for the UK to grow and thrive and I welcome this book as a significant contribution to the goal of wider application of new technologies.

Alan Carnell
President, EEA.

Foreword

'A Guide to Advanced Manufacturing In Electronics' has been prepared as part of a joint Industry and Department of Trade & Industry project on AMIE. Project members formed four working groups with the following tasks:

PG1: To prepare a guide book on AMIE and make recommendations for awareness activities.

PG2: To study the achievements of a number of home and overseas companies in the use of AMIE, and determine if lessons already learned could be applied more widely in the UK.

PG3: To examine the current status of AMIE standards and identify areas in which action is needed.

PG4: To examine the needs of smaller electronics companies for AMIE.

The knowledge gained in these activities was made available to the author, and PG1 provided an editorial panel. The following individuals served on the panel:

R. W. J. Ward	ICL Network Systems
M. G. Ginn	GEC Research Ltd
T. J. Weare	Pactel (formerly of Schlumberger)
L. Ashton-Smith	Schlumberger Measurement & Control Ltd.
J. H. Brownlee	Schlumberger Measurement & Control Ltd.
G. Blyth	Marconi Instruments Ltd.
B. Hunt	CAP Industry Ltd.

The Department of Trade & Industry wishes to acknowledge the substantial contribution made by the panel.

Chapter 1

Advanced manufacturing in electronics

The book discusses the application, integration and management of technology in the design and production of electronic products.

1.1 Introduction

'The application of advanced manufacturing in electronics will be a key influence in determining the future competitiveness of the UK electronics industry.' This was the view of Cameron Low, Chairman of the Electronic Capital Equipment Economic Development Committee in a recent NEDC discussion paper.

Companies which have taken hold of the competitive advantages offered by well planned and implemented applications of Advanced Manufacturing In Electronics (AMIE) illustrate the truth of the above statement. They have seized opportunities presented by world markets which double in size every three to four years. But why AMIE? Indeed what is AMIE? What can AMIE provide that a more traditional approach to the design and manufacture of electronics products does not?

In short, AMIE can provide significant marketing advantages. This is particularly apparent in terms of quality and responsiveness—factors which today's increasingly sophisticated customers are ranking as being major differentiating factors in supplier selection. AMIE also gives the customary cost-saving benefits associated with automation. Table 1.1 summarises the range of benefits already being experienced by users of advanced automation. Furthermore, the use of advanced-components technology opens new marketing horizons by enabling a considerably enhanced or radically different product.

Examples of companies which are leading the field are certainly not restricted to large foreign companies. For instance, a small UK company manufacturing pocket-sized cellular telephones has expanded its market by exploiting advanced technology. Making use of fully automatic assembly methods for a complex and tightly packed design, they have produced a reliable high-quality product with a unique and superior specification and enhanced performance. The costs involved in designing the product and setting up the manufacturing facility have not been small. However, investment was not limited by restricting its justification to traditional cost saving. Management had a broad view of the competitive advan-

Table 1.1 *AMIE benefits*

Engineering design cost	Down by 15—30%
Overall lead time	Down by 30—60%
Quality level, yield	Up by 200—500%
Assembly productivity	Up by 40—70%
WIP level	Down by 30—60%
Organisational efficiency	Up by 5—20%

tages available to them through the use of both high and low technology. They assessed the risks of standing still and they also assessed both the risk and benefits of moving forward. The opportunity for benefit arising from AMIE was too great to be missed and many advantages were offered—all vital to any company wishing to succeed. The benefits included:

- Higher quality and reliability of product and service—lower rework and warranty costs, improved market image.
- Shorter design to market availability time—a larger sales window and increased market share.
- Reduced tendering time—improved response to customer.
- Decreased production cycle time—more flexibility, improved response to customer.
- Reduced unit cost—improved margins, more sales.
- Improved product specification—in functions, performance, size, weight, power consumption.
- Increased efficiency—more output (design and product) for same resources.
- Better asset management—improved cash flow.

Little explanation is needed for such benefits, but in the volatile electronics market, where explosive developments occur frequently, they must assume a place of considerable importance. It is not uncommon to discover a product being technically obsolete before it reaches the market, or that development time cycles now exceed the sales windows. The market demands an increasing proliferation of special products, and unless appropriate action is taken, development cost and time escalate astronomically.

It is against this background that the properly planned implementation of advanced manufacturing techniques can provide the ability to meet these exacting needs. However, the basic issues must be faced up to and resolved. Sustained improvements must start with top–down planning. This is a strategic approach, built upon long-term business goals, reflecting the needs of the market place rather than merely considering and applying technology in a piecemeal form to resolve isolated problems as they emerge.

The need to couple technology to strategic planning is well illustrated by a further example. This is again drawn from the UK electronics industry, and is a major manufacturer producing communications equipment for both home and world markets.[3] Poor management and a fragmented approach to business organisation and operations resulted in a pass rate at first-time test of under 30%. No one really knew the exact figures nor the underlying reasons, and neither was any action being taken to improve the situation. Other shortcomings compounded the unprofitability. The plant was swamped with work in progress (WIP) and delivery schedules were never achieved. Investment had not been stinted, but technology had been introduced for its own sake rather than to resolve clearly identified business problems or as part of an overall business strategy.

Planning cannot be carried out in an insular fashion, considering only the UK market and domestic competitive forces. There are few companies which will not be either facing the problem of foreign imports or wanting to exploit overseas markets. Generalisations are often meaningless in the context of individual companies; nevertheless some salient factors are evident when considering the competition arising from Japan and the USA:

- Non consumer goods from Japan often lag behind their Western counterparts in specification.
- A frequent comment made by visitors to Japan is 'there is nothing here which we couldn't do at home.'
- Spending on design for manufacture is at a very low level in the UK as compared with Japan.
- Performance of many US companies equals that of their Japanese counterparts.

Factors such as these should inspire optimism within the UK electronics industry. Indeed, even within the consumer-goods sector, radical improvements have recently been achieved. The obverse of optimism is complacency. This may arise from many sources, including traditional British attitudes or from a false sense of security experienced as a result of feeling safe in current niche markets. Complacency kills: long-term viability can only be enhanced by realism now and strategic investment in technology, organisation and skills.

This book is deliberately aimed at presenting practical solutions in order to resolve today's problems, whilst building foundations for tomorrow's increased profitability and viability. The following Chapters detail the vital stages of planning and implementation, illustrating the benefits of AMIE and outlining the basic elements of AMIE such as those illustrated in Fig. 1.1.

1.2 What is AMIE?

This book is a guide to AMIE. It is not, however, a comprehensive description of the technology, and can give only a very cursory view of even the most significant techniques. Rather, it is intended to put technology in perspective by illustrating the benefits which AMIE can provide when implemented in a well planned, structured and strategic fashion. Few readers will need to be told that technology is rarely a simple plug-in solution. Indeed, as the definition of AMIE illustrates, it is as much to do with methodology and management as it is with technology itself.

This book develops the benefits which various approaches can realise. A key stage in the initial planning stages is identifying the area likely to provide the greatest return on investment. The correlation between areas of application and benefits will depend upon individual company circumstances; nevertheless typical relationships are given in Table 1.2.

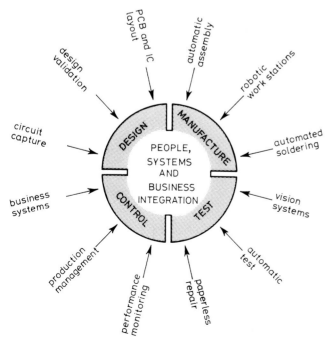

Fig. 1.1 *AMIE building blocks*

1.2.1 Technology

AMIE consists, at a basic level, of a number of building blocks concerned with the tasks of cost-effectively and competitively designing and producing finished electronic goods. As Fig. 1.1 shows, these are split into the four main functions of:

- Design
- Manufacture
- Test
- Control

Within each of these major functions there will be a number of activities. Fig. 1.1 illustrates some of the possible options. For example, automatic test splits very roughly into three further subdivisions; in-circuit, functional and emulation. Design validation broadly divides into logic simulation, timing verification, analogue simulation and thermal analysis etc. The exact combination required in a given situation will obviously depend upon the nature of the individual business.

There are two important aspects of AMIE which will be common to almost every implementation programme:

- Design carried out on CAD, ideally making use of well thought out component data bases and design validation techniques in order to make 'right first time' a reality.
- Performance monitoring—enabling efficiency, yields and quality problems to be analysed as a means of providing corrective feedback on a real-time basis.

The forerunner to AMIE was CADMAT—Computer Aided Design, Manufacture and Test. CADMAT was concerned with the application of technology to the tasks of design, manufacture and test. Invariably this resulted in the now infamous 'islands of automation', or perhaps better termed 'seas of separation', where individual functions are automated in an insular fashion. AMIE advances considerably beyond CADMAT in that it not only automates the individual functions within a company, which will often include the business systems, but it also integrates these processes. Genuine integration will ensure that information generation, transfer and exchange are as automatic as necessary.

AMIE cannot end with technology. As we can see from Fig. 1.1. integration is as much

Table 1.2 Business impact by AMIE application

AMIE application	Improving quality and reliability of product and service	Material, labour and overhead cost reduction	Ability to introduce new products sooner	Ability to cope with greater variety	Improved product specification and performance	Ability to reduce and achieve production time cycles
Just-in-time/organisation	●	●	●	●	·	●
Computer-aided design	●	●	●	●	●	●
Manufacturing automation	●	●	●	●	●	●
Automatic test	●	●	●	●	●	●
Production management	●	●	●	●	·	●
System integration	●	●	●	●	·	●
Component technology	●	●	●	●	●	·

Note: Significance of impact denoted by marker size

applicable to people as it is to systems. The integration of business functions is also a key aspect of the implementation of AMIE. For example, there is little value in having a CAD system if the requirements for manufacture are not built in to it, or having a manufacturing requirement planning package if component purchasing lead times are not taken into account through a lack of communication and multidisciplinary co-operation. Integration of both equipment and functions is central to the concept of AMIE.

Effective implementation of AMIE can only be achieved if based on a secure foundation of sound business-oriented goals and strategic planning. Much of this planning will be to do with analysing how the business operates. Once the business has been understood, it can be treated as a process capable of being controlled effectively. Thus, although the correct choice of technology is important, it will never be sufficient in its own right, and the way in which AMIE is implemented and subsequently managed is also vital.

1.2.2 Components

Component technology cannot be separated from AMIE, and this book will consider the relevance of Application Specific Integrated Circuits (ASICs) and other component technologies. Innovations and developments in component technology have had a dramatic impact upon the cost, reliability and specification of products. Of particular importance today is semi-custom IC technology. This is now within the reach of very small companies and those companies having low individual product volumes. Several key messages are emerging as being relevant to this technology:

- Application is becoming widespread
- Start-up investment is modest
- Commercial benefit can be immense.

Unless note is taken of such developments, there is a great danger of being left behind with an uncompetitive product in the midst of an increasingly discerning market place.

1.2.3 Related approaches

AMIE is concerned with electronics products. However, advanced manufacturing as a whole embraces many other technologies, which have varying degrees of application within electronics. Of the plethora of acronyms and 'buzz words' that are prevalent today, three are worthy of special mention:

- Computer-integrated manufacture (CIM)
- Flexible manufacturing systems (FMS)
- Just-in-time (JIT).

There are many definitions and interpretations of these terms, but the following discussion puts them into the context of AMIE. More detailed descriptions and examples of specific applications can readily be found in documents such as the proceedings of related annual conferences.[4]

Computer-integrated manufacture: Information in the right place and at the right time to enable a task to be performed correctly.

Computer integration forms the technical core of this. The definition presented is particularly apt because it is not restricted to asking such questions as 'how many robots?' It is concerned with ensuring that any process or function is presented with information in the correct format to enable a task to be carried out at the appropriate time.

The utilisation of expensive capital plant and scarce, highly skilled staff are both often restricted by a lack of appropriate or correct information. Costly and error-prone manual data-transcription tasks must be eliminated if utilisation is to be optimised. Computer integration can make a substantial inroad into improving business efficiency in such areas.

Computer integration seeks to make use of the most appropriate computer tool for a particular task, and helps to ensure efficiency by integration. Implementation of computer integration will vary according to the specific nature of each company. For some it will include large main-frame systems with massive centralised data bases, while for others it may consist of a network of microcomputers with relatively small amounts of on-line data storage. Either system can be appropriate in the right environment.

Flexible manufacturing systems: A system which combines microelectronics and mechanical engineering in order to bring economies of scale to batch work. A computer controls the work stations and the transfer of materials, components and tooling. It also provides monitoring and information control.

A flexible manufacturing system is a specific application of computer-integrated manufacture. The broad definition of FMS given above often implies a high degree of dexterity in mechanical manipulators in order to achieve flexibility. However, as the example given in the second company mentioned in Chapter 10 illustrates, a high degree of automation is not always necessary in order fully to utilise FMS.

Flexible assembly cells or systems are a particular subset of FMS, and the significance of these is likely to increase as technology advances. Systems such as the one illustrated in Fig. 1.2 are becoming available, providing the capability at a PCB level of achieving the purpose of FMS defined above.

FMS by definition does not include the design function. Nevertheless FMS cannot be divorced from design. If any system is to succeed, the manufacturing constraints must be incorporated into the engineering data base, and in this respect FMS is no different from any other approach.

Undoubtedly a flexible manufacturing system will meet the objectives of some companies. However, an FMS is not an instant package that can be overlaid onto existing practices. A preliminary to any investment is a thorough appraisal of the business. This will reveal the value of, for example, investment, re-organisation, education, training and new methods. This can result in a solution which will provide far greater benefits than a stereotyped FMS package, which may only serve to automate existing inefficiency and malpractice.

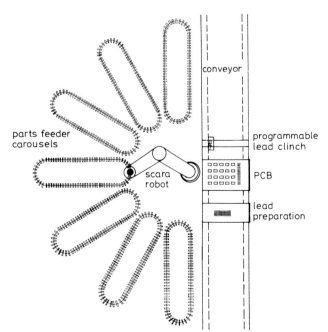

Fig. 1.2 *Flexible assembly cell*

Just-in-time: An applied philosophy in which an activity is initiated just-in-time to meet the output requirement.

The concept of JIT was developed by the Japanese automobile industry in the early 1970s with three key objectives:

- Reduce inventory
- Eliminate poor quality
- Reduce manufacturing costs.

These are, however, somewhat at variance with a prevalent Western image of Japanese JIT, where it is perceived to be mainly concerned with deliveries arriving every hour. JIT is concerned with internal efficiency, organisation and practices. Although supplier interaction is important, JIT also seeks an uninterrupted flow within the factory. Key aspects will include implementing total quality control, removing bottlenecks which cause WIP, simplifying work flows and reducing queues and inventory.

Furthermore, the application of JIT is not restricted to major companies with leverage over their suppliers. For example, a company producing electronic products for machine tools reduced the lead time from 44 to 3 days, and a company producing oscilloscopes achieved a similar improvement. In another instance, the quantity of suppliers has been reduced by a factor of 20, from 640 to 32, with a corresponding increase in the quality of incoming material and supplier service, and a very substantial decrease in procurement costs.

More details of JIT are given in Chapter 7. This discusses in more depth the approach to the technique and its benefits.

1.3 Summary

A challenge is presented to every company by growing markets. The increase in the availability of technology and knowledge enables companies to respond and thus exploit those markets. However, it would be wrong to presume that AMIE, or indeed any solution biased towards any particular technology, will be the complete answer. A balance is required—matched to a company's own situation, carefully assessed, meticulously planned and effectively managed.

This book is a guide to the tasks facing management. It presents experience gained over a number of years by representative users in a wide range of companies. It is intended to point others in the right direction and to the first rung of the ladder. Although it covers many aspects of AMIE it can never be fully comprehensive, and the reader is urged to use it as a guide—to promote thought, investigation, planning action and profit.

1.4 Bibliography

1. GILLAN, R.: 'The Japanese secret—Are they winning'
2. GREGORY, G.: 'Japanese electronics technology, enterprise and innovation' (Wiley, 1986)
3. 'Winning with AMIE' (National Economic Development Office, 1987)
4. Proceedings of the International Conference on Flexible Manufacturing System (IFS (Publications) UK, 1982 *et al.*)

Chapter 2

Justification

This chapter discusses the benefits and costs of AMIE, outlining the methods for financial justification and the tools available for the task.

2.1 Introduction

Economic justification of investment in technology is always difficult, particularly when evaluating advanced technology. In the early days of automation, whether it was for automatic component insertion or computer-aided draughting, the main savings were normally associated with direct labour. As technology has advanced and has become more complex and far reaching, this traditional and often very narrow view of cost saving may fail to demonstrate the overall viability of an investment. Investment in new technology has an impact throughout the company, and is usually not restricted to the commissioning department. For example, the ability of CAD to produce 'right-first-time' designs, and to generate assembly and test programs, can create a greater benefit outside the design department than it does with the department itself.

Many of the benefits arising from investment in AMIE have often been classed as intangible and unquantifiable. Investment may take place following intuitive attempts at justification based on the so-called intangibles. However, in many instances, investment is impeded by the inability to generate a fully quantified case, even though many of the so-called intangibles are in fact capable of quantification.[3] In the extreme, technology may be pushed to its limits in order to provide direct cost savings, while a lower-technology, and hence lower-cost, solution could have given a greater proportional benefit.

This chapter will review the benefits, both tangible and intangible, and the costs of AMIE. Some examples of the approaches to justification taken by companies will be given, together with a consideration of justification techniques.

2.2 Benefits of AMIE

Benefits will arise in many parts of the company from the implementation of AMIE. This section discusses the benefits of the type illustrated in Fig. 2.1 which ought to be realised in a well planned and managed implementation. The monetary benefit will obviously depend on the installation, and

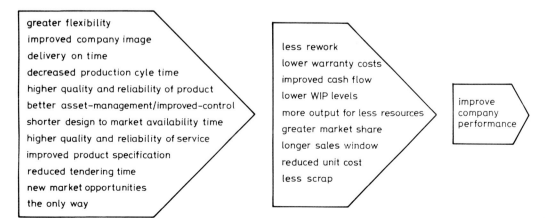

Fig. 2.1 *Gaining the benefits*

precise quantification is therefore outside of the scope of this section. However, in Fig. 1.1 an indication is given of a range of benefits which could reasonably be expected. Benefit will not only come from investment in technology itself, but also from a radically different approach to organisation and an attempt to eliminate problems at source. These additional factors are discussed in Chapter 7.

2.2.1 Higher quality and reliability of product

The use of design automation, improved manufacturing techniques, quality assurance and a greater thoroughness of test helps to ensure that products are of a higher quality and more reliable. Design simulation can identify and eliminate tolerance problems which would otherwise be missed by the physical approach to design. Company standards can be maintained through the use of an engineering data base. The elimination of manual assembly techniques will decrease reliability problems associated with defective soldering and static damage. Use of advanced component technology will provide a more reliable product through the reduction of interconnections in the form of wiring harnesses, backplanes and printed circuit boards. Finally, the provision of accurate statistics on actual yield and quality helps to allow electronics production to be treated as a process. This is capable of real-time control and results in a substantially lower defect rate than with a batch-type manufacturing operation.

Improved quality and reliability of product will have major benefits:

- Reduced scrap
- Reduced re-work
- Reduced warranty costs
- Improved marketing image.

2.2.2 Higher quality and reliability of service

Quality of service, whether it be to an external customer or to an internal department, is vitally important to any organisation wishing to remain competitive. AMIE can serve to improve the quality of service in a number of ways. Accuracy and stability of schedules are improved through the use of automation, which helps to perform tasks 'right first time' and thus eliminate re-work. This is not restricted to design, but applies equally in production, where automation can also provide a product which is 'right first time'.

On a broader level, the use of planning and scheduling tools, computer integration and the implementation of JIT techniques can all help to ensure the product being delivered on time to the customer. For example, on-line access to up-to-date and accurate computer records can enable an immediate, accurate and effective response to be given to a customer, instead of an inadequate reply through reference to out-of-date and thus unreliable hard-copy information.

Thus market image will be enhanced through the on-time delivery of a higher-quality and reliable product.

2.2.3 Reduced tendering time

A much improved customer response and supplier image can be achieved by a reduction in tendering time. Although the sales function is not considered in this book, computer-aided production management and expert systems can help to make this activity more systematic and responsive. In a sample of six companies implementing advanced manufacturing, the resulting reduction in tendering time varied from $10:1$ to $5:1$. The quality of tender documentation can also be improved through the use of CAD. Preliminary work carried out during the tendering stage can be carried through into design. Once again, an improved customer image and response can result, together with the possibility of a reduced product development time.

2.2.4 Shorter design/market-availability time

The correct use of design-automation tools, coupled with automatic assembly, can radically reduce the lead time from new product concept to market availability. A company producing specialised equipment to order will be able to provide shorter deliveries, thus improving its competitive position. Internally, shorter project time scales will enable a job to be costed and controlled more accurately.

Companies involved in mass production must also be able to grasp the opportunities provided by developments in component technology by incorporating these into the product at the best time to gain a competitive edge. Computer-based design tools described in Chapter 12 are starting to allow 'what if' type questions, thus allowing conceptual designs to be evaluated and the optimum design to be achieved.

A shorter design to market time provides for

increased sales, a longer sales window and a greater market share.

2.2.5 Decreased production cycle time

Decreases in production cycle times give the two very significant benefits of improved flexibility and additional responsiveness. Improved flexibility means that resources need not be committed until the last moment. Thus, for example, products can now be built to order instead of for stock. The latter approach suffers from the dangers of stock write-off or of material stock-outs whilst the urgent production of orders is denied the material being consumed by speculative production.

Decreasing the production cycle time can also have a profound impact on quality, re-work and changes. For example, consider companies A and B. These companies each have a radically different approach to their business, resulting in production cycle times of six weeks and 24 hours, respectively. Many unnecessary production problems are created and compounded by the elongated production times endured by company A, whilst company B, without these problems, is able to concentrate its resources on effective production.

Company A struggles with poor soldering quality. This is a direct result of oxidisation and contamination of component leads and bare PCBs. The company has to allocate considerable resources to the task of work tracking in order to be able to deal with shortages and changes, to expedite urgent orders and to help give some semblance of production control. The company is constantly implementing a multitude of changes to in-process product, and is building to complex change notes rather than to simple and definitive manufacturing documentation. Finally, the company endures a very high level of re-work because production defects are not identified and limited by effective in-process control mechanisms.

Within company B the situation is very different. In-process inspection rapidly detects production defects and the line is stopped before any appreciable quantity of defective product has been made. The very short production time cycle virtually eliminates the problems associated with the implementation of design-originated changes to in-cycle production. Shorter cycle times and the minimum exposure of leads and bare PCBs to the atmosphere eliminate oxidisation and contamination problems.

The benefit of decreased production cycle time is not limited to the shopfloor; benefit is also seen in management areas. Management is now released from the firefighting tasks of implementing changes and dealing with problems arising from high defect rates, and can now actively manage. In addition, computer systems are reduced in complexity because detailed shopfloor work tracking is now no longer required.

Benefits can therefore be achieved in customer response, improved utilisation of resources and lower direct and indirect production costs.

2.2.6 Direct cost savings

Automation can provide some readily quantifiable benefits over many manual methods. This is seen not only in manufacture, but also in design and other support activities within a company.

Reduced unit cost: In addition to the direct cost savings associated with automation, there are unit costs reductions arising from other sources, such as the reduction in production and material costs. Use of the appropriate technology can achieve a reduction in raw material cost and the cost of assembly. This is well illustrated by one of the case studies in Chapter 10 where a 60:1 reduction in assembly time is cited. Enclosures can be smaller and lighter, with fewer parts. Heat dissipation may be less and power supply requirements reduced. The reality of such benefits are, for example, reflected in the costs of engineering work stations which are approximately halving every 18 months or so. In addition, each new model shows a marked increase in performance and a reduction in physical size over its predecessor. Reduction in unit costs can provide higher margins and also lead to increases in sales and market share.

Improved product specification: There is an incessant growth in the demand for technology. It is fuelled by awareness of the possibilities of technology itself, by the drive for 'bigger and better' and for greater business efficiency. These demands create an environment where there is a constant requirement for products with an improved specification which offer better value for money.

The utilisation of advanced component technology, particularly in the form of ultra- and very-large-scale-integrated circuits (ULSI/VLSI) and application specific integrated circuits (ASICs), creates the opportunity to develop radically different products. Depending upon the nature of the product and the market, specification improvements in the areas of function, speed, performance, size, weight and power consumption all contribute to opportunities for increased sales and profits and for exploiting new markets.

Increased operating efficiency: Increased operating efficiency can provide either more output from the same resources in people, floor space and equipment, or it can provide a cheaper product.

The provision of computerised tools and the integration of systems allow more output, in design or production, without an equivalent increase in staff. This has a further benefit of maintaining such overhead costs as accommodation, services and management, whilst increasing output.

The use of automatic assembly and the reduction in material storage and handling requirements will allow a more efficient utilisation of space. This results either in more output from a given floor area, or a reduction in the factory space required for a given output, where a saving of 50% is not unusual.

Increased operating efficiency of equipment can stem from many sources. Integration of equipment, elimination of defective production and taking positive action to reduce set-up time, all help to increase the throughput on the shop floor. Integration, which allows assembly and test programs to be automatically generated from a CAD system, is of particular relevance. This eliminates on-line programming and the de-bugging of manually introduced program errors, and releases costly equipment for production.

Such improvements in efficiency all contribute to an increase in the return on investment and the ability to manufacture a greater volume of product more cost effectively.

Increased management effectiveness: The most significant improvement in efficiency may be seen in the area of management. Better information systems, which provide immediate and accurate data, are essential to effective decision making. Management can be released from firefighting and holding postmortems which enables them to carry out their prime task of managing and directing the business.

Better asset management: Changes in operating procedures, the introduction of JIT techniques and the management information available from computer-integrated systems all help to improve the management of a company's assets. The resulting decrease in WIP, increase in stock turns, decrease of stock write-off, reduction in production scrap and control of production all help to improve cash management in a company. This can, in turn, reduce funding charges or release capital for further investment.

2.2.7 Improved company image

Investment in high technology can do much to improve the image of a company, both externally and internally. Customers rightly associate CAD, auto-assembly, ATE and bright clean modern facilities with efficiency and quality. Internally, this investment can help boost employee morale, provide stimulation and motivation, and help to reinforce the commitment of staff to the future of the company, while assisting the recruitment of the right calibre of staff.

2.2.8 New market opportunities

Use of advanced component technology, provision of enhanced design and production capability can all help to create new market opportunities.

2.2.9 The only way

As technology advances and customer requirements become more exacting, it is evident that the application of AMIE techniques may present the only way of carrying out a task. For example, there is no option but to use computer aids for the design of custom integrated circuits. Surface mount technology is moving to a point where automatic assembly will be the only available option. Hand soldering carries such a penalty in long-term unreliability that, even for very low manufacturing volumes, this method may become totally unacceptable.

2.2.10 Summary of benefits

The benefits arising from effective implementation of AMIE are manifold and, as Fig. 2.1

illustrates, they can all help to increase sales and promote company growth and profit. Many of these benefits can readily be quantified. Some, such as the value of increased sales resulting from an improved market image, will be more difficult to assess, but, although many are indirect, very few are really intangible.

For example, marketing management should be encouraged to establish the parameters of change. They could be asked to establish the increase in sales resulting from a 10% cost reduction or from the introduction of a product which is half the weight of its predecessor. Production could be asked to establish the effect on stock write-off of an increase in stock turn of, say, three to eight per annum. Quantification of the variables is made easier by these simple cause-and-effect questions.

2.3 Justification case studies

The following two case studies are included as illustrative examples of both the approach which two different companies took to justification, and also the types of benefits which were realised for implementations of AMIE.

Case study 1
Implementation of CAD included the derivation of the basic ATE programs directly from the CAD. Three main savings were identified in the area of ATE programming. These were:

- Basic test program design, such as the allocation of test points and channels, could be eliminated.
- There would be no data entry of information such as net lists, parts lists or board layouts.
- Manually introduced errors in data entry and fixture build could be eliminated, thus reducing the on-line debug of programs and fixture.

ATE program developments took six weeks per new design. It was estimated that the integration of CAD and ATE would save three weeks per new design, split between the above three activities in the ratio of 1:1:2. There would be a saving of approximately 35% in the cost of developing an ATE program for each new design.

A further potential benefit was also identified. Additional ATE capacity was required to cater for company growth. The increased throughput resulting from the implementation of this integration link would reduce the requirement for additional equipment.

Case study 2
One manufacturer produced large PCBs, containing mainly connectors and costly custom VLSI devices. Up to 200 components, from a range of 70, were required to be inserted into the 16-layer PCB, measuring about 20 × 15 in.

Quality of assembly was of paramount importance for several reasons:

- Devices were prone to static damage.
- Devices were very costly at up to £100 each.
- Re-work was difficult with the non-standard 179-lead VLSI devices.
- The PCB was expensive.
- A number of combinations of product were to be assembled using the same-sized PCB.

The nature of the product precluded the use of conventional auto-insertion equipment. Manual assembly was an option, but, because of the associated quality problems, it could be expensive.

Finally, a robotic solution was considered. Cost savings were identified, resulting not only from a very substantial reduction in scrap and re-work but also from eliminating assembly documentation, component preforming and other associated back-up tasks. Furthermore, the assembly system was linked to the CAD to enable the automatic creation and down-loading of assembly programs. In the justification procedure the system throughput was considered. Compared to normal auto-insertion equipment the robotic solution was slow with a 6 second cycle time. However, quality was of much greater importance than the rate of throughput, which could be achieved through parallel lines and an elongated working day.

A payback period of two years resulted from a £200 000 investment in a flexible automation system.

2.4 The cost of AMIE

The justification of AMIE is related to potential benefits and costs. It is important that all the costs involved are analysed, including both the once-off and recurring costs.

Once-off costs are usually split into the following main categories:

- Capital cost less any grants
- Consultancy costs, less any grants
- Feasibility and planning costs, less any grants
- Non-capital installation costs
- Training costs
- Recruitment/redundancy costs
- Accommodation and expenses.

The success of a project depends on careful and exhaustive planning. The cost of this can be substantial and should not be underestimated. The levels of grant currently available for this task indicate the sort of commitment that will be required for even a medium-sized project.

Finally the people costs associated with implementation can be substantial. These will include recruitment and redundancy costs, the cost of training and expenses during evaluation and training periods.

Once the system is installed and running, the costs continue:

- Direct labour
- Labour shift premium
- Supervision
- Maintenance, including software and people
- Consumables
- Energy
- Insurance
- Support activities
- Tooling
- Financing charges or depreciation.

Excluding direct labour, maintenance may be one of the most significant of all the running costs. For many computer-based systems an indicative 'rule of thumb' for maintenance cost is 10% of the current combined hardware and software list prices. Internal support activities are associated with maintenance costs.

Computer systems do not run themselves, and often a system manager is needed together with some ancilliary support staff at various levels. Requirements for facilities usually increase as users discover the benefit of computer integration. Development of these facilities using a high-level programming language may be carried out internally. Data bases also require maintenance in the form of routine clearing of redundant information and updating with new or additional information.

Systems will often have to be operated for extended periods if the required return is to be obtained; so premium labour rates may have to be paid.

2.5 Financial evaluation methods

2.5.1 Payback methods

For many engineers the already difficult task of justification is exacerbated by a lack of understanding of the financial computations themselves. Consequently, a very simplistic view is often taken of payback period and rate of return. For example, an investment of £40 000 produces savings of £16 000 per annum to give a payback period of 2½ years. However the return also depends on the working life of the equipment and its scrap value. Assuming a 10-year life and 10% scrap value, this will, by reference to Fig. 2.2, give a rate of return at current prices of about 40%.

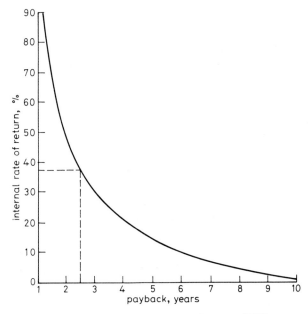

Fig. 2.2 *Payback vs. internal rate of return (IRR)*
16% scrap value
Constant return in each year

Such a simplistic approach has a number of shortcomings, the most important of which are that it ignores the time value of money and that a two- or three-year payback period, as shown in Fig. 2.2, demands a return of 33—40%. This compares very unfavourably with the normal level of 15—20%, which is more often the accepted range for other investments.

2.5.2 Discounting techniques

Investment in AMIE will invariably be phased and complex in nature. It often spreads over several years from the start of investigation through to completion of a major project. Taxation and capital grants, which are also time dependent, will also affect the payback. For example, taxation payments and repayments are normally delayed by one or more years after the liability has become due, and capital grants are invariably paid retrospectively. Furthermore, finance for major projects may well have to be obtained from external sources such as bankers or investors. The case presented to them must be both financially sound and presented in a fashion they can readily understand. The appraisal method must meet their requirements. It must be able to estimate and measure performance in a way that takes account of project life, the pattern of profit generation, taxation rates and residual value. Discounted cash flow techniques (DCF), as opposed to simple payback methods, account for the time value of money and allow all the quantifiable factors to be explicitly analysed and compared on a common basis. A formalised technique such as this will also allow the sensitivity of a project to be analysed against variations in assumptions and risks. Relative profitabilities of different project structures or solutions can also be compared.

There are two ways in which financial appraisal using discounted cash flow techniques may be used—these being net present value (NPV) and internal rate of return (IRR). The latter is often preferred by non-financial managers because the concept is similar to the one used for repayment mortgages. The NPV method is preferred by many accountants and is, as the name implies, concerned with the value of a project at a particular point in time.

Internal rate of return: Rate of return at which discounted future cash flow equals the initial cash outlay.

This method is best illustrated by an example. Suppose that, in order to finance a new project a company borrows £40 000 at a rate of 15%. The project earns £24 600 in each of the next two years. The annual earnings could be viewed as a capital repayment of £20 000 and profit of £4600. Cash flow for this project is set out in Table 2.1, example A. The return on this project is 15%. In other words, the investment has both recovered the capital and paid interest, or return, on the investment of 15%.

However, assuming the same capital positions at the start and finish, a change in timing could have a substantial effect on the project, as example B in Table 2.1 illustrates. In this instance, the investment generates only £8000 in the first year but £40 000 in the second; the rate of return is now only 12%.

Although the total earned in each example is identical at £49 200, example B has a lower rate of return because the earnings arise later in the project and consequently more loan is outstanding at the end of the first year.

Net present value: The as yet unrealised capital gain from going ahead with a project.

This method attempts, as the name implies, to evaluate the net present value of a project at any point in time. This could be equated to the price at which one might wish to sell the project to a third party. In this method the annual net cash flows of a project are discounted at a chosen discount rate. The difference between the total of the present value of cash

Table 2.1 *Discounted cash flow—internal rate of return*

		Example A £		Example B £	
Year 1	Loan	(40 000)		(40 000)	
	Interest	(6000)	15% of 40 000	(4800)	12% of 40 000
	Paid back	24 600		8800	
Year 2	Loan outstanding	(21 400)		(36 000)	
	Interest	(3200)	15% of 21 400	(4400)	12% of 36 000
	Paid back	24 600		40 400	
	Balance	0		0	

14 Justification

Table 2.2 *Comparison of methods—examples*

	Project capital cost, £	Annual Capital Recovery					
		Year 1	Year 2	Year 3	Year 4	Year 5	Year 6
Example A	40 000	10 000	10 000	10 000	10 000	10 000	10 000
Example B	40 000	16 000	16 000	4000	4000	10 000	10 000
Example C	40 000	10 000	10 000	10 000	10 000	20 000	20 000

inflow and total initial cost is described as being the net present value.

2.5.3 Comparison of methods

For some projects a simple approach based on payback may be sufficient. However, the question has to be asked as to what is an acceptable payback period? Furthermore, the total economics of a project may be radically altered by what happens after the payback period is completed. Table 2.2 gives three further examples; all of these have a payback period of four years. Using the payback method it is impossible to choose between them. However, because of the higher capital recovery in years 1 and 2, example B appears preferable to A or C. Applying DCF techniques gives a radically different result. Despite the favourable initial conditions, project C is now preferred, with an internal rate of return of 20% as opposed to 15% for B and 13% for A.

2.5.4 Sensitivity analysis

Sensitivity analysis allows the investment appraisal to be tested against variations in parameters. For example, a range of answers will be given in reply to questions such as 'by how much will sales increase for a 10% reduction in selling price?', or 'what reduction in warranty cost will arise from a 20% increase in reliability?' Any appraisal should be 'tested' with the range of answers in order to ascertain how critical that parameter is.

Each strategy option should be subject to sensitivity analysis. The parameters which will be tested will vary depending on the project, and typically they will include: the effect of a longer time scale, variations in labour cost, change in interest rates, raw-material cost variations, volume throughput, wage settlements and taxation laws. Their outcomes can be expressed in the form of, for example, 'a 15% increase in capital cost would reduce the return by 2%'. This provides authoritative information on the basis of which judgements on the choice of options can be made.

2.5.5 Risk analysis

Risk analysis takes sensitivity analysis a step further. Risk analysis seeks to establish which option has the greatest chance of succeeding. The probability of achieving the most likely result is estimated for each element analysed for sensitivity—for example, that there is an 80% certainty of meeting time scales or 70% of gaining the increased market share.

Having established these figures, the management team can agree a weighting, or importance, against each factor. Alternative strategies can be assessed by examining their respective weighted probability of success, which is, as Table 2.3 illustrates, the sum of all the individual success probabilities and weightings; some of these values can be negative, as the Table shows. Risk analysis is an aid to assessing which strategy is most likely to succeed, and therefore which one should be followed through. The overall outcome can be calculated in financial terms and expressed as internal rate of return or net present value evaluations.

2.6 Tools for financial evaluation

Having established the appropriate approach to the problem of financial evaluation, there remains the task of carrying it out. Analysis can proceed, as Fig. 2.3 shows, only after the costs and benefits have been detailed. Many of these figures can be readily obtained; for example, the capital cost of purchase, the decrease in WIP level and the increase in productivity can all have fairly accurate values assigned to them. Indirect benefits often require a more detailed investigation in order to establish the basic data and consequent estimates. This may well involve market

Table 2.3 Risk analysis

Factor	Weighting	Strategy option A		Strategy option B	
		Probability	Score	Probability	Score
Increased sales	90%	0·7	+63	+0·5	+45
Increased capital cost	80%	−0·5	−40	+0·1	+8
Project escalation	40%	−0·2	−8	−0·5	−20
		Total	15		33

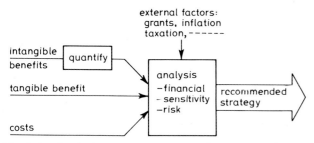

Fig. 2.3 Analysis procedure

research to establish market sizes, price elasticity and additional opportunities.

In this initial quantification stage care must be taken to ensure that thorough and detailed analysis is not influenced by cautiousness or over-optimism. Whilst being conservative in estimating may prevent unsound projects from advancing, too great a caution may limit further investment, thus depriving the company of benefits. Given the correct input and the appropriate approach, analysis can now proceed. As already discussed, discounting methods are often chosen. At a basic level all discounting methods rely on establishing what a given sum of money will be worth at a particular point in the future. Discount factors are the reciprocal of compound interest factors. For example, if £1000 earns interest at 10% compound for two years, its value at the end of two years will have become £1210. Similarily the discount factor for £1000 costing 10% compound over two years is 1000/1210 or 0·826. Such factors are tabulated in discount tables and can readily be established using calculators or simple computer programs.

An added refinement will be to use a spreadsheet approach. Once having created the appropriate model with the various inputs, this allows the analysis to be rapidly and accurately carried out. Sensitivity and risk analysis will both also be produced as a result of parameter variation. Whilst the use of a spreadsheet is a considerable advance over manual techniques, it does entail building the initial model. Work has been carried out in order to develop a formal and standard approach, suited to managers, engineers and accountants. For example, such tools may present output information in the form of tables, charts or histograms, giving projected cash flows, rates of return and net present values.[6,11]

As with all other computerised evaluation tools, this also gives the ability to ask 'what if?' questions, evaluate the options and highlight the possible areas of risk. It can therefore be an aid to establishing areas for further evaluation. Such a tool must not be used blindly. It is important that the company has a thorough understanding of the basic operation of the program and the assumptions which it makes. Any assumptions must be carefully checked against company practices and performance, and also against current taxation and grant regulations.

2.7 Summary

Appraisal is a vital prelude to any investment. Whilst intuitive feelings may be correct, they can never be a substitute for thorough analysis. Intuition can be the basis for sanctioning investment in, for example, CAD, but appraisal will provide the direction in which to proceed in terms of system capacity and the way in which integration should take place.

Again considering CAD, if only direct savings are taken into account, investment in CAD may be restricted to low-cost stand-alone systems. A more thorough and comprehensive analysis, including the benefits in production and ATE, could well indicate that a substantially more expensive system would be a very much more sound investment. The quality and thoroughness of the analysis will also help to assess the 'intangibles', and thereby improve the investment decision.

Financial evaluation can now be aided by computerised tools. The tools available now are only a prelude to future developments. The integration of expert systems, artificial intelligence, market models and so forth will provide a comprehensive market-led analysis procedure.

Investment analysis is still only an aid to management decision making. Whilst computer-based tools can help to make a sound decision rapidly, they are no substitute for a sound and thorough basic approach where the indirect benefits are identified and quantified.

2.8 Bibliography

1. CREAMER, G. D.: 'The evaluation and selection of CAD systems'. Proceedings of 7th Design Engineering Conference
2. DONALDSON, W. M.: 'AMT—Selecting the most profitable option'. Proceedings of CADCAM 86
3. PRIMROSE, P. L.: 'Evaluating the 'intangible benefits of FMS by the use of discounted cash flow algorithms within a comprehensive computer program', *Proc. of I.Mech.E*, **199** (B1)
4. PRIMROSE, P. L.: 'The use of financial viability to determine the future of AMT'. Proceedings of CAD-CAM 86
5. PRIMROSE, P. L.: 'The use of simulation data to determine the optimum economic application of FMS'. Proceedings of 1st International Conference on Simulation in Manufacturing
6. SHEEHAN, V.: 'Financial evaluation of capital projects' (Organisational Development Ltd.)
7. SMITH, W. A.: 'A guide to CADCAM' (Institution of Production Engineers, ISBN 0 85510 029 X)
8. STEVENS, P. R.: 'Investment appraisal', *Production Eng.*, June 1971
9. YU, W. S.: 'Case study: Design of an FMS for producing TV tuners'. Proceedings of FMS 4, 1985
10. 'Advanced manufacturing technology—The impact of new technology on engineering batch production' (National Economic Development Office, ISBN 07292 0704 8)
11. 'Computer model user handbook' (P-E Information Systems Ltd.)

Chapter 3

Design

This Chapter discusses the basic design process. It outlines the major computer-aided tools available, including those for circuit capture, simulation and layout. The result of the design process is information. Methods for using this as the source of manufacturing and test instructions are considered.

3.1 Introduction

The emphasis of design activities is changing, with design for manufacture now being a key objective. This is in order to reduce the time and cost of bringing new products into the market place. Designers have been preoccupied with ensuring the correct functionality of a circuit and minimising component cost.[3] They are now much more aware of the requirement to design for manufacture in order not only to reduce the manufacturing effort, and hence cost, but also to design with automatic and high-yield assembly in mind.

The growing scope and power of CAD systems allows a radically different approach to design. CAD extends the boundaries of design into the traditional domains of laboratory breadboards and prototypes, assembly and test.[1] Furthermore, design is increasingly being viewed as an information producing function, providing the majority of all manufacturing and test information, such as photo plots, drill tapes, parts lists, auto-insertion programs and test programs.

This chapter examines the design process and describes the available tools, their scope and possibilities and the hardware on which they run.

3.2 Design process

The traditional, and usually manual, approach to design is reflected in Fig. 3.1. The seemingly straightforward route of translating a requirement specification into a circuit, and eventually into a reliable product, is a lengthy and tortuous iterative process, full of pitfalls and difficulties. Errors creep in at many levels as information is manually copied between documents. Breadboards cannot readily show the effect of parametric variations such as component tolerance. Time invariably forces engineers into design release before the circuit is fully evaluated. The volume of documents associated with a single design imposes a formidable burden on drawing-office staff in terms of creating, checking and updating.

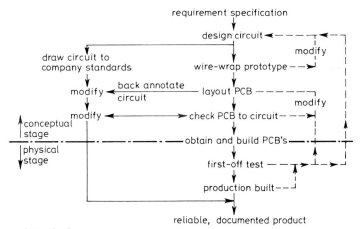

Fig. 3.1 *Traditional approach to design*

18 Design

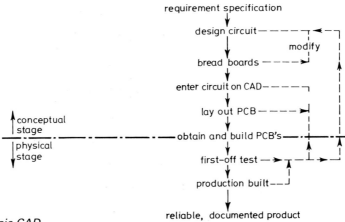

Fig. 3.2 *Design using basic CAD*

Finally, because of the inadequacies of such a process, many design problems are not identified until production or pre-production units are manufactured and tested. A plethora of design changes results.

Inevitably the results are unacceptably long design cycles and excessive manufacturing costs. Defects identified during production prove to be very costly, owing to additional design and drawing office work, scrapping or modification of products and the obsolescence of pre-purchased material. However, of potentially much more importance are the competitive disadvantages such as vital new products being delayed by many months whilst problems are identified and resolved.

It is against such a background that the well managed introduction of basic CAD can resolve many of the difficulties. Circuit entry and board layout capabilities of these systems can make dramatic reductions in design and manufacturing costs, cut time scales and eliminate many design modifications. Such a development is reflected in Fig. 3.2 where the main improvements arise in such areas as:

Circuit capture, where libraries ensure that only approved components are used, that information such as pin numbers or signal names are correct and that basic electrical design rules such as gate loadings or interconnection rules are adhered to.

PCB, hybrid circuit and IC layout, where component libraries, standard outlines and design-rule checking programs guarantee the adherence to manufacturing and operational requirements, such as component spacing and orientation, clearances, track widths, pad and hole sizes etc.

Documentation management which is considerably eased with back annotation and checking facilities eliminating vast quantities of manual effort. Data bases, with their associated management, ensure the control of documentation and adherence to design release procedures.

The effect of these tools can be dramatic. It is not unusual to find that the introduction of basic CAD can cut the number of PCB design iterations by three, that the design cycle time is cut by 30–40% and that auto-assembly can be fully exploited because the requirements for it are built into the CAD database.

Full CAD, shown in Fig. 3.3, where, in addition to the basic facilities, software simulation of the design is also carried out, should lead to 'right-first-time' designs where any design iterations crossing the conceptual/physical boundary are eliminated. Simulation programs enable the designer to simulate the functionality of a circuit, evaluate the effect of parametric variations and also carry out testability analysis. Once a product has been physically designed, these programs can be used to check that effects such as track or inter-track capacitance and cross-talk do not materially affect the operation of the circuit. Some systems also provide other tools such as thermal modelling and vibration analysis. Mechanical CAD systems may be integrated with some electronic CAD systems in order to provide physical tolerance checking of a PCB within an enclosure and the transfer of common mechanical information.

CAD brings major benefit to multi-unit designs, whether it be a single PCB wired into an enclosure or a major system comprising custom ICs and a large number of PCBs. The CAD tools facilitate data management, re-use

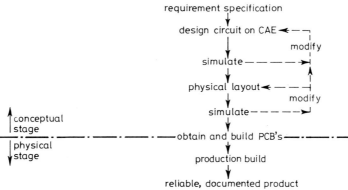

Fig. 3.3 *Design using full CAD*

of data, cross checking of information between units, and system-level design verification and simulation. These provide an unprecedented opportunity for ensuring the accuracy of such designs and cutting development times and costs.

CAD readily lends itself to those products which require dedicated test equipment. In particular, military products traditionally require the correct correlation of both product and bespoke test apparatus—a task which usually proves to be a substantial overhead and problem. The complete system can be designed as a single entity, with the effect of changes in the product being evaluated against the test equipment design and the two aspects kept in step.

Once a design has been captured the power of the CAD system allows trade-off analyses to be carried out, considering, for example, aspects of cost in relation to component specifications or circuit partitioning.

Documentation is one of the final stages of design. Data held within the CAD system should be used to drastically cut the effort involved in the task of producing manufacturing, maintenance and operational documentation.

Design cannot be thought of as an isolated process. Not only is design for manufacture important, but also the CAD tools produce information which can be used as the source for the majority of all assembly and test information. Traditionally CAD output has been seen as a file to drive a pen or photoplotter. However, as far as the CAD system is concerned, there is in essence no difference between, for example, a plotter and a component insertion machine, since both require information in digital format about specific aspects of the design. If the benefits of automation are to be fully realised, integration is vital and the role of CAD must be viewed more widely as providing the information required to assemble and test an item in addition to actually designing it. The automated design environment shown in Fig. 3.4 illustrates the key features in terms of both information and facilities offered. These are developed further below.

Finally, CAD implementation is concerned with more than just hardware and software. The manner in which it is managed is vital and governs its effectiveness. For example, if manufacturability is to be guaranteed, the manufacturing requirements must be built into the data base, design release and modification control procedures require to be carefully designed for the specific environment, and users need training in the wider aspects of design.

3.3 Circuit capture

Until recently many users or potential users have viewed schematic entry as being almost optional, with the prime purpose of CAD being to lay out a physical product. Falling hardware prices have enabled a greater proportion of users to see this phase as an integral part of the whole. The alternative to schematic entry is to describe a circuit using component and connection net lists. Whilst this may allow a rapid entry of information, the subsequent checking of layout to circuit and any back annotation or modification for even a moderate sized design becomes an unwelcome, time-consuming and error-prone task.

A variety of data entry methods are possible. For example, with PCB-based designs the prime method of data entry is using conventional circuit diagram symbols. For semi- and full-custom integrated circuits there are often options available in the form of

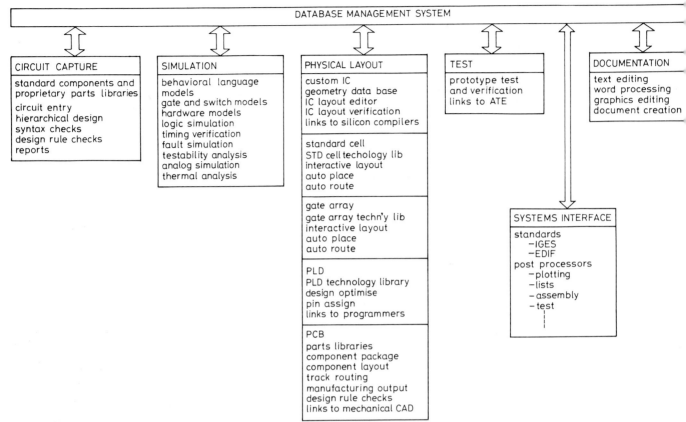

Fig. 3.4 *Automated design environment*

Boolean and behavioural or functional descriptions of a circuit. Software will automatically decompose this into primitive circuit elements related to the target component technology.

Systems invariably offer a multi-sheet schematic facility whereby a circuit may be composed of several sheets. Few systems now fail to provide a hierarchical approach whereby a circuit may be created either from a block diagram level down, or detailed circuit diagram level up. Hierarchy may extend to many levels, providing the ability to create not only a detailed description of a custom IC, but also the PCB on which it is based and the system within which that PCB resides.

This approach, with connections being made by signal names as well as visible point-to-point connections, lends itself to major systems using many discrete modules and designed by large teams of people. All documentation can now be readily cross-checked and any errors highlighted while still of minor significance. A natural spin-off benefit is the provision of wiring information for system racks and panels.

Depending on the system, design partitioning may not be carried out until after simulation has taken place. In this process the decision is made as to how the circuit design will be split up, or partitioned, in terms of PCBs, hybrid circuits, ICs etc. This allows the circuit designers to concentrate initially on the functionality of the product without undue concern about the details of manufacture or even the component technology that will finally be used.

For PCB layout the allocation of logic gates to physical packages may come at this point, or, depending on the system, may not occur until the PCB layout stage. In either instance the system should facilitate management of the gates and components used, and provide an automatic method of ensuring that the final circuit has the correct pin and package information. This process is known as back annotation.

CAD systems vie with each other for user friendliness, and offer many options for speeding up both the entry of electrical data and information concerned with the human understanding and aesthetics of the circuit. Key features are often related to part naming, connection entry, data buses, pin- and signal-name incrementing, rubber banding, and circuit copying and moving.

Circuit entry is not only concerned with capturing information; it is also about ensuring the quality of that information. Once verified, information contained in a component library will be correct. Therefore every time a component is used, whether it be a two-leaded polarity-conscious device or a 128-pin IC, pin numbers and names will be correct and the company standard drawing symbol used. Furthermore, few systems do not provide checking facilities which ensure that the circuit is electrically correct. This facility will warn of such potential problems as technology mixes, gate overloads and short circuits.

As a final stage of circuit entry there are invariably checking and reporting options which provide, for example, listings of signal names, components, unused pins, single pin nets etc., with all items cross referenced to sheet and zone. These will rapidly highlight minor data entry errors such as calling a data line DATO and DAT0 in different places.

Reliability calculations can be carried out at this stage, using static characteristics. If, however, the effect of electrical and thermal stress on components requires evaluation, the appropriate simulation tools should be used at a later stage.

The circuit capture phase will make available three important outputs:

- Documentation that is well presented, clear to read and understand, and which meets company standards.
- Information in digital format for the next stage of design.
- Information which is electrically correct.

3.4 Simulation

Although the circuit capture phase should have provided information which obeys basic electrical design rules, there is no guarantee of circuit functionality or that parametrically the circuit is within specification. This is the task of simulation. As integration increases in importance the role of simulation expands to include the interaction with product test. This is at two levels: it ensures the testability of the design and also provides test programs in the form of stimulation and response data. Associated fault simulation information is also provided to aid the fault finding task.

The available tools cover many different aspects of simulation, but the following are the main categories:

- Analogue
- Logic
- Fault
- Timing
- Thermal.

The scope and power of these tools continues to rise as CAD systems develop. Designers are offered an ever increasing ability to ensure that a design is 'right first time'.

3.4.1 Analogue simulation

Analogue simulation has been a 'poor relation' in this field, with little more available than fairly basic tools for verifying transfer functions of linear circuits. Recent developments have started to provide software design environments which can not only emulate an analogue circuit designer's laboratory, but provide statistical analyses and calculate reliability data related to the electrical stress on individual components. Comprehensive and powerful interactive graphics provide a very user-friendly environment in which the designer can operate.

Software tools provide the functionality of signal generators and oscilloscopes, voltmeters and power sources, frequency sweepers and spectrum analysers. In addition to this basic instrumentation, are control functions which allow the effect of component tolerance, power supply variation, temperature and the like to be evaluated and analysed, often using Monte Carlo type statistical analysis. Component sensitivity analysis permits the tolerance requirements for components to be thoroughly evaluated, with tolerance-critical components being high-lighted. Power dissipation calculations, both mean and peak, allow component ratings to be checked and semiconductor safe operation conditions to be verified.

These developments, when integrated into the overall design environment, provide a very powerful alternative to the laboratory, and eliminate laborious and time-consuming manual calculations.

3.4.2 Logic simulation

Logic simulation provides the software design environment in which the functionality of a digital circuit can be verified. Primarily, the user generates an input pattern which may be expressed in a variety of ways, such as a timing diagram, truth tables or a stimulus file. The simulator evaluates the effect of this input pattern on the logic circuit, using information about the functionality of the devices obtained from a software model. In the absence of a software model, some tools allow the use of the physical device itself.

An accurate and comprehensive library is essential. Whilst most simulators come with a library, newer, and usually complex, devices may not yet have a model. The writing of a model is a complex and lengthy process, and therefore many users will derive the model from a real physical device. Functionality is derived from this, whereas timing information is derived from manufacturers' data sheets.

The output of the simulator can be viewed in a number of ways, such as timing diagrams or truth tables. Data can be manipulated, for example, by comparing the expected with actual simulation results. The effect of a design change can be assessed, with differences from previous simulations being highlighted.

Simulators operate in a variety of modes. Some work in a batch mode and provide interactive graphical facilities to examine and manipulate the data. Others provide an integrated facility, thus allowing rapid interaction. However, the volume of data produced in a simulation can be very large, and therefore some simulators will store only the simulated patterns for specified circuit nodes. Although this may ease data storage requirements and improve speed of simulation, it can create delays whilst re-simulation is carried out if different circuit nodes have to be examined.

Some simulators are event driven whilst others are not. An event driven simulator will only evaluate those parts of a circuit which are affected by an event. Thus, although an input may change, the whole of the circuit may not have to be evaluated as with other techniques. This reduces the computing power required to carry out the simulation task compared to alternative simulation techniques.

Design simulation is directly related to product test. The design simulation process verifies the operation of the complete circuit. In essence, it tests the circuit. The stimulation and response patterns can be used as the basis of functional test programs, thus considerably reducing ATE programming time.

Interaction with the product is also increasing the use of 'hardware verifiers', which allow the physical system to be connected to the simulator and the real operation compared with the simulated. In a very low volume environment this process may also be used as the actual production test facility.

Finally, the simulator is a very powerful tool for identifying the effect of component variations. A simulator will clearly indicate whether the logic state of a device can be deduced from known information. In a normal breadboarding environment the output of a device will depend on random conditions, e.g. individual component characteristics or power supplies. Thus the prototype circuit may appear to work correctly. However, once the spread of component tolerance and other parameter variations are experienced, very costly and time-consuming problems are likely to arise. The logic simulator is a powerful tool for not only verifying correct operation, but also identifying areas of potential trouble.

3.4.3 Fault simulation
In this process the logic simulation of digital circuits is carried a step further to evaluate the effect of faults on the operation of a circuit. Typically, each node is subjected to three faults; namely, open-circuit output, node stuck at zero and node stuck at one. The simulation process is repeated for each fault. Two outputs will emerge: a list of any faults which cannot be identified and an associated percentage fault coverage. Additionally, diagnostic information relating the particular fault to the impact on simulation output is determined. These are important, as they give confidence in the degree of design verification that has been carried out and provide information to aid the diagnosis of faults at test.

3.4.4 Timing
Logic simulation of digital circuits is usually carried out using standard gate delays. Timing verification allows the effect of propagation delay and other time related parameters to be evaluated, helping to highlight glitches and race conditions that may occur in practice. As with other forms of simulation the effect of component tolerance spread can be evaluated whilst the design is still at the conceptual stage, and therefore whilst the cost and time involved in correcting the design are still relatively low.

Different approaches are taken to timing verification in order to decrease the computing power and time required. These usually involve the simulator only evaluating timing variations when actually required. For example, if a parallel circuit exists, then only one of the two routes may be time critical and therefore only that would be evaluated.

The initial logic and associated timing verification will initially only use data concerning the conceptual design. However, once the design has been physically created, parameters such as capacitance will change, producing a consequential effect upon delay and switching times. Simulators are often integrated into the physical layout tools to evaluate the impact of layout, accounting for aspects such as track capacitance or cross talk. Some of the more advanced simulators will deal with multilayer as well as two-sided PCBs.

3.4.5 Thermal

Thermal simulators often form a part of the layout system, and their use comes after a product has been physically designed. This simulator will allow the designer to check the power dissipation and distribution across a design. Heat dissipation by radiation, conduction and convection should be able to be simulated, including heat sinks or forced methods such as air blowing. A product-temperature profile will be created to verify operations within the specified temperature range and to examine the effects of fan positions, air-flow rates and heat-sink geometries.

3.5 Layout

The PCB layout process consists of two distinct, but related, major stages. These are component placing and track routing. Whilst auto-routers, and their percentage routing, are emphasised by many vendors, the more important phase is component placement. Unless the placement is very carefully carried out and with the subsequent stage of routing in mind, the routing process will be exceedingly difficult, if not impossible, to complete. As component density increases, so this task becomes more critical.

While every system will require a slightly different strategy for PCB layout, the processes described below are typical of the majority. Some more sophisticated systems employ a radically different approach to those described. Several systems exist for hybrid-circuit layout which use algorithms suited to the application and produce masks for each individual layer.

3.5.1 Component placement

Output from the circuit entry phase will primarily consist of a list of circuit elements and their interconnections. This is transferred into the PCB design environment and combined with any default PCB design data such as board dimensions, positions of standard features—holes, tooling, panel layouts etc.— and design rules—clearances, pad and hole dimensions etc.

Depending upon the system, the circuit elements may already be packaged into physical components or this task may form one of the first activities of placement. Once packaged, auto-placement software will attempt to place each component on a grid position, obeying rules about layer usage where surface-mount devices (SMDs) are concerned. The components placed at this stage are usually ICs, with other components being left to a later stage.

The final position of components should not be random. Optimal placing will usually be achieved using automatic and interactive facilities. Invariably there will be some manual interaction in the form of pre-placement of some areas, zoning of the design, or movement of auto-placed devices.

Placing must be carried out with routing in mind. Although minimum connection length is important, other factors need to be taken into account. Careful consideration needs to be given to the placing of components connected to data buses. In particular, memory arrays require devices to be in-line to facilitate the daisy-chaining of tracks. Interconnected components should be grouped together, e.g., placing external bus drivers close to their associated connector. Invariably the placing can be improved by exchanging equivalent gates between packages and equivalent pins within a gate. This is usually an automatic feature. The number of available routing channels should be considered to ensure that there will be sufficient room to run all the tracks. There are often automatic features, such as connection-density histograms, to help check this. These provide a clear indication of potential difficulties in routing, through lack of routing channels.

Placement of discrete components usually follows IC placement. This task is usually carried out totally interactively unless there are a large number of components to be placed on a regular grid.

The exact sequence of placement will vary according to the nature of the design. The above method is typical for a digital-based circuit where auto-placement software is extensively used. Analogue-type circuits are often placed totally interactively, with auto-placement being only used at a later stage and if there is any associated digital circuitry.

Surface mount technology imposes peculiar requirements upon placement. For many applications a component may be placed on either side of the PCB. In order to ensure flexibility, the system should be able to place a component on either side of the PCB, mirroring the connection-pad pattern and component outline as appropriate.

Component placement is a key stage in design for manufacture. Placement is the stage at which manufacturing requirements should be incorporated. If automatic assembly is to be used effectively the design must incorporate the particular requirements of each item of equipment. Typically, insertion footprints and component orientations must be catered for. Building in the manufacturing requirements must not be viewed as restricting the designer, but as a help to the designer to achieve a quality design which can be manufactured reliably and cost effectively.

3.5.2 Track routing

Track routing follows placement. Routing is usually achieved with a combination of interactive and automatic tools. The exact sequence of operation will depend on the type of circuit. An analogue circuit may be routed entirely interactively, whilst a two-sided digital board will probably be routed using a combination of interactive and automatic routing passes.

Auto-routers fall into many different categories, but the most common usually utilise two main strategies: a memory router and a random-search-pattern approach.

The memory router will attempt to route connections which fall within a strict definition of length and orthogonality. They will usually be up to about 1·5 in long and will not deviate from the orthogonal by more than, say, 0·1 in. Such connections are deemed to be memory type, as would be expected if there were a row of memory packages. Routes should include angled portions to enable maximum routing efficiency to be achieved.

The random-search-pattern router attempts to put in routes by looking for routing channels and utilising these as appropriate. Allowing an auto-router to run indiscriminately can give poor routing-channel utilisation. Control of the router is therefore necessary. A first auto-routing pass may limit either the computing power which can be used to attempt a route, or can put restraints on the connections that may be attempted. For example, a first pass may limit the orthogonal deviation of connections, and only attempt virtually horizontal or vertical ones. After this pass, short connections with a slightly greater angle would be routed. Routing parameters would be gradually relaxed until the most complex connection was tried. The reason behind such an approach is to stop a complex connection being put in first and blocking many routing channels, thus preventing simple connections being routed.

Rip-up routers route connections until a channel is blocked by a previously routed connection. This track will be 'ripped up' and the new connection attempted. If the system deems that the new solution is better than the old, it will be retained or else the old is restored. These routers may rip up many existing tracks in order to try a new one.

A multi-pass router will attempt to route the PCB a number of times, but each time using different routing parameters. Thus it could be left to run in batch mode and the user presented with the solutions when it had completed operation.

Auto-routers which work to fixed rules rarely achieve a final solution which is as good as a manually produced one, and therefore final connections are often better routed interactively. Interactive modification of auto-routed tracks may be necessary in order to create additional routing channels.

An important feature of CAD is design-rule checking. This may be on-line to prevent the user from violating them, or it may be a post-routing facility. Track layout can also be checked against the circuit.

Much mechanical information is available from the PCB layout stage in the form of hole sizes and positions, board profiles and component positions. Systems may provide an integrated PCB and mechanical CAD environment, or data interchange will allow mechanical draughting to be carried out on a separate CADCAM system.

3.6 Documentation

The generally unwelcome task of documentation is considerably eased by those systems which provide documentation facilities. These may allow word processing to be merged with graphics to provide a straightforward method of annotating and describing the operation of circuits and products. A common feature is to have documentation as a part of the design data base to enable, for example, any occurrence of a circuit update to be automatically revised within the documentation.

3.7 Post-processing

The CAD system will have produced detailed and accurate information concerning the product during the various design phases. This will include data concerning what components are used, how they are connected, physically placed, routed, and information about how the circuit actually works. All this information is held within the engineering data base, but usually in an outwardly unintelligible format. It is the task of post-processing to convert this into useful manufacturing and test information.

There is considerable advantage to be gained by using CAD-produced data. This arises from not only being able to produce programs more rapidly, but also by reducing the debug time which would otherwise be required from manually introduced data-transcription errors. There is a spin-off benefit in allowing increased equipment utilisation, by eliminating on-line programming effort and by the reduction in debug time.

Neutral data formats, such as EDIF, are increasingly gaining acceptance as their benefit is realised and more vendors utilise them. In essence, these formats make the source and target systems independent of each other and ease the problems involved in software maintenance as systems are updated. This is covered in more depth in Chapter 8.

3.8 Libraries and data bases

Systems require libraries which describe individual components and data bases which describe the products. Component libraries contain all the information associated with a component to allow it to be used by the CAD system. This would include all the draughting, electrical, functional, physical and textual data associated with that part. The integrity of the data is vitally important because the accuracy and reliability of the design, in every respect, will depend on it. Standard information as contained in a data book—electrical and functional characteristics, physical size, timing data, tolerance and the like—often comes direct from the CAD supplier in a purchased library. Company-specific information, such as circuit diagram symbol, stock reference, auto-insertion footprint, orientation restrictions, pad and hole dimensions etc., will require definition by the user. This is no mean task, and is often carried out progressively as new designs are carried out.

The integrity of this information is important. Data-base management is vital and users must not be allowed indiscriminately to update it. However, management procedures must cater for a rapid response to user update requests if users are not to be unduly restricted and delayed. Procedures should be established to allow the forewarning of impending new devices, either to enable them to be on the system by the time they are required or to warn the user that the device will not be allowed, why it will not be and what the alternative is.

Product design data-bases are equally important. Design-release procedures are required to ensure that designs have gone through the required verification and checking stages in order to ensure that they meet the specified standards.

Many types of design environment and data-base structure exist. Some of the more common are outlined in Figs. 3.5–3.7. At the most basic level (Fig. 3.5), discrete data files hold different categories of information. Different application programs access these, and yet further programs translate the data between them. This creates undesirable overheads as resources are channelled into performing these tasks and in ensuring the correlation of different files rather than in pure design work. Generally, many lower-cost PC-based systems utilise this approach, as will systems comprising tools from a number of vendors.

Moving up a stage, a single relational database holds all the design information associated with a design, as Fig. 3.6 shows.

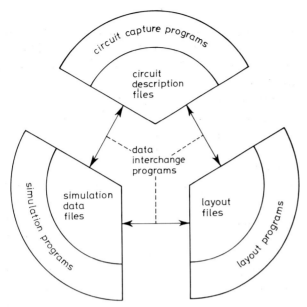

Fig. 3.5 *Fragmented system structure*

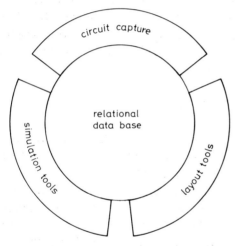

Fig. 3.6 *Conventional relational data-base approach*

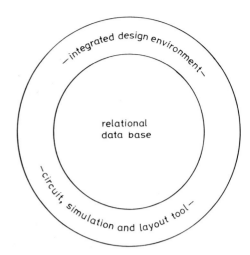

Fig. 3.7 *Integrated design environment*

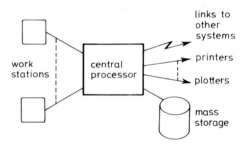

Fig. 3.8 *Centralised processing system*

There are now no programs required to transfer information between, say, the circuit entry and layout programs. The design-tool programs are, however, different. If, for example, during the simulation phase the circuit requires modification, the simulator must be left and the circuit-entry package evoked to implement the required change. This type of system, as marketed by several of the major CAD vendors, is easier to manage and use than the fragmented approach outlined in Fig. 3.5.

At the top level, as shown in Figure 3.7, a totally integrated environment is provided to facilitate concurrency between the various design activities. Thus an activity of the type outlined above could be carried out very rapidly and without leaving any of the application packages.

3.9 Hardware

There are two main types of hardware configuration for multi-work-station systems, these being based upon either centralised or distributed processing as illustrated in Figs. 3.8 and 3.9. Centralised processing may utilise intelligent terminals to remove some of the graphics-processing load from the main processor.

Distributed processing hardware may range from networked low-cost personal computers through to very powerful work stations. The network can include shared resources such as computing engines, file servers, hardware modellers and verifiers, together with peripheral devices like plotters and printers. Falling hardware costs are changing the configuration of systems. For example, the use of low cost PCs for data entry with intelligent work-station-based systems is decreasing as work-station and PC costs and technologies converge.

The distributed approach is often to be preferred as it can be more easily expanded without a loss in performance, as Fig. 3.10 illustrates. One system has well in excess of 100 nodes with no loss in performance. However, network design is important if response to shared resources is not to prove to be a limiting factor.

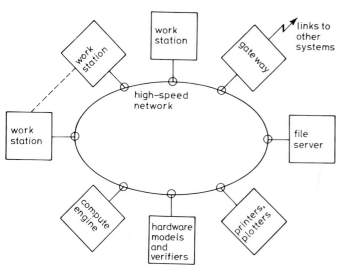

Fig. 3.9 *Distributed processing system*

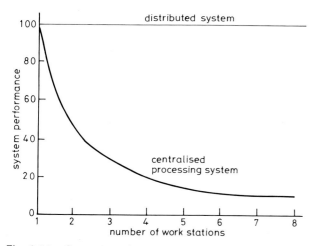

Fig. 3.10 *System performance*

for check plots and wet ink for customer documentation. The penalty is in speed and entity representation and an electrostatic plotter may be better. Laser plotters, which combine very high speed with quality, are becoming increasingly common.[2]

A key output from CAD is photo-plotted artwork. For small users photo-plotting is an expensive overhead and it is therefore often performed by a bureau. The most common type of photo-plotter is a flat-bed unit, while scanning types, utilising a modulated laser beam, are becoming more common. A number of users achieve the required accuracy by photographing images printed on high-density dot-matrix printers. This option is not recommended for large boards, or where fine lines and spaces exist. For low-volume or ultra-precision work, artwork may be dispensed with entirely by photo-plotting directly onto photo-sensitised copper laminate. This technique is growing in acceptance as it eliminates many of the quality and accuracy problems associated with artwork.

3.11 Summary

As we have seen, CAD needs to be viewed very widely. It is more than just an automated drawing board, and should have significant impact throughout the company and not just in the design office. The future of CAD lies in systems with increased performance and facilities, offering the efficient design of both volume products and one-offs. Designs which can be cost effectively manufactured will become the order of the day as design and production problems are pre-empted by the use of standard information, extensive simulation and, most important of all, systems which ensure that designs are carried out according to the correct rules and procedures.

3.12 References

1. McGREGOR, J.: 'Analysing manufacturability and the effects of design changes', *Printed Circuit Design*, May 1986
2. MARTIN, P.: 'Laser photoplotting as an integral part of CIM,' *New Electronics*, **20** (6)
3. 'Designing high density PCBs cost effectively' (Du Pont de Nemours and Co., Inc., 1982)

Irrespective of the configuration, effective data management is vital to prevent duplicate copies, data divergence and to have secure archive and back-up.

3.10 Plotting

Hard-copy information will be required from the system for a variety of purposes ranging from check plots through to customer documentation. Large systems will often have a variety of devices—dot-matrix printers, electrostatic plotters and pen plotters—while a small user will often be restricted to a single device. For these users, a pen plotter can provide flexibility in terms of colour capability

Chapter 4

Manufacture

This chapter discusses the major stages of PCB assembly, covering component insertion and placement, soldering and cleaning. It also outlines some of the possible approaches to materials storage and handling.

4.1 Introduction

A wide variety of equipment is available to assemble components onto PCBs. This ranges from high-speed volume-oriented equipment through to simple aids for manual assembly. Although in its infancy, robotics is increasing in importance.

The great diversity of individual company operational characteristics—costs, volumes, varieties, component types and mixes, product specifications, work-force skills and design methods—all make it impossible to give detailed specific parameters for equipment choice. The main equipment approaches for different volumes and mixes are shown in Fig. 4.1.

There is considerable variety between different companies in the way in which they apply assembly techniques. For example, one company auto-inserts well in excess of 90% of all components, while another company, manufacturing essentially identical products, considers that automation limits production flexibility, and consequently auto-insertion is used for less than 30% of all components. In another instance, a supplier of auto-insertion equipment has supplied equipment to companies with an annual insertion requirement of 500 000 components, but, along with other suppliers, has failed to supply equipment to companies with 10 times that insertion requirement.

Irrespective of the specific type of equipment used, the nature of the product will dictate the exact production processes. The three main routes are illustrated in Fig. 4.2. Product containing only leaded components will follow a route based upon Fig. 4.2a. Parts of this will be optional; for example, good design can remove the requirement for post-flow assembly with the associated task of solder-mask application.

A different approach is required where SMDs are used. Mechanical location and retention of components during production is no longer provided by component leads. Two routes are possible: in the first, illustrated in Fig. 4.2b, the SMDs are glued into position on the PCB prior to flow soldering; in the second,

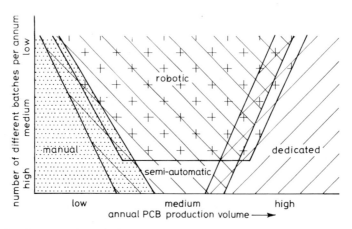

Fig. 4.1 *Automation application by volume/mix*

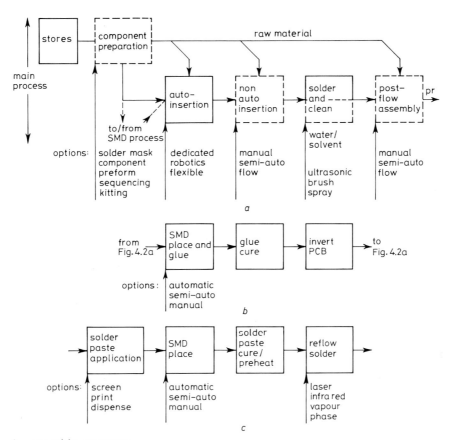

Fig. 4.2 *Electronics asembly sequence*
 a Leaded component process flow
 b SMD placement-flow solder
 c SMD placement and re-flow solder
 Note Optional processes shown dashed

illustrated in Fig. 4.2c, thick and tacky solder paste is used to hold components in position during assembly. Re-flow soldering melts the solder paste in order to provide the permanent electrical connections.

Mixed-technology product, i.e. product with both leaded and SMD components, will generally follow the route shown in Figs. 4.2b and 4.2a. Pure SMD product will follow that shown in Fig. 4.2c. A complication arises when SMDs are placed on the same side of the PCB as leaded components. A two-stage soldering process is required and the assembly route would be a concatenation of all the processes shown in Figs. 4.2c, b and a, respectively.

4.2 Assembly methods

4.2.1 Manual assembly

The superior human abilities of creativity, judgment, flexibility, visual discrimination and fine control of motion make the manual process an attractive potential option for many assembly tasks. However, for all but very small total volume applications, this method will suffer from problems associated with a low speed and poor intrinsic quality. The latter results from electrical and mechanical damage to components, incorrect components being inserted, mis-insertion and poor lead crop and clinching giving rise to solder short circuits. Manual assembly is therefore usually reserved for the insertion of a very limited number of odd-form components once the scope of automatic equipment is exhausted.

4.2.2 Semi-automatic

In order to improve both the quality and speed of assembly, a number of different types of semi-automatic equipment have been developed. In the main these now utilise a software programmable indicator light, which, under computer control, is co-ordinated with a parts-dispensing mechanism. The indicator light will guide the operator to the target position for a particular component, indicating polarity when important, and possibly giving some limited message to the operator via a small display. Some machines provide a mechanised lead crop and clinch mechanism in order to secure components.

This type of equipment obviates the requirement for any learning specific to a particular board type. It provides considerable quality and throughput improvements in either low-volume, or the initial stages of high-volume, production. However, these machines often work quite slowly and are rarely used for volume.

4.2.3 Flow line

Flow line assembly is often a logical progression from pure manual methods as volumes start to increase. Product 'flows' from one work station to the next. Various layout options are possible, two of which are illustrated in Fig. 4.3. At each work station a limited number of parts are inserted, usually working against a master or marked-up drawing. Intrinsic quality tends to be higher than with pure manual assembly. Flow line methods are often used in a volume environment for the insertion of odd-form components following dedicated component insertion. An interesting example of flow-line assembly is cited in the second case study in Chapter 10.

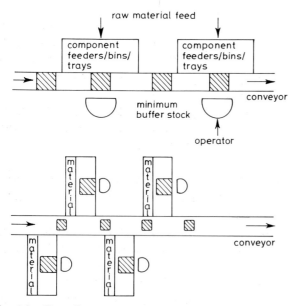

Fig. 4.3 Flow-line assembly

4.2.4 Medium-volume/batch equipment

This type of automatic equipment is specifically aimed at manufacturers making medium volumes of product, usually in small- to medium-size batches and with a wide variety of components. It can find application in companies producing fewer than 20 000 PCBs per annum. Features of such equipment are rapid programming and set-up, manual board load and unload insertion rates in the range of 900–3000 components per hour and medium cost—generally less than £60 000.

Components are selected from one of a number of parts feeders and automatically pre-formed prior to insertion. The component styles which can be catered for by a single machine vary with equipment type. Some can cater for only a single component style, whilst others can insert axial, radial and dual-in-line components.

Some of these machines can accept a pre-sequenced bandolier of components, whilst others allow a battery of component feeders to be replaced at once. Such facilities permit rapid changeover from one product type to another. A further useful technique is re-firing. Re-firing allows, in the event of an assembly problem being detected, for the component to be discarded, another selected and assembly tried again. This option can provide substantial quality improvements with, for example, the yield of one particular machine increasing from 98% to 99·5% when this option was used.

4.2.5 Dedicated equipment

Very high-speed insertion equipment is available which will insert components at rates of several tens of thousands of components per hour. Production speed is usually gained at the expense of limited flexibility with elongated change-over time between products.

Invariably such equipment will provide automatic board handling. Manufacturers of insertion equipment often provide integration facilities to allow both the physical linking of equipment and the electrical connection into host controllers. Physical integration may cater for alternative or parallel routes for certain sections of the process to facilitate line balancing.

4.2.6 Surface mount device placement

Surface mounted device placing can employ similar techniques to those used for leaded components. However, there are some significant differences. A single machine can be used to place all SMDs, whereas with leaded components a different machine is usually required to insert axials, radials, dual-in-lines and some odd forms. Thus many of the line-balancing problems associated with leaded-component insertion are eliminated, with a commensurate reduction in set-up tasks and production control and management. Furthermore, smaller components may allow many more items to be held on-line on one machine, thus providing for greater production flexibility. For similar performance the SMD equipment tends to be considerably cheaper and smaller than comparable leaded equipment. This can have an important impact when considering the factory space requirement, and cost, required in order to achieve a given output.

However SMDs do bring problems. There are substantial differences between SMDs and leaded components. The majority of small SMDs have no marking to indicate component type or value. Electrically identical devices may be physically different. This imposes very severe restraints on the use of manual assembly. Even with semi-automatic methods the danger of mixing components is far too great for many users to accept. The increasing range of automatic equipment available, coupled with the emerging packaging standards, is increasingly making automatic placement the only viable option.

Components require holding in place either by solder paste or glue. Solder paste is usually applied by a screening process, although for small-volume assembly, robotic-controlled dispensing is a viable option, whilst glue is usually applied by the placement machine.

4.2.7 Robotics

Within the UK the use of robots for assembly purposes is still at an embrionic stage with few cost-effective solutions yet in existence. This situation is slowly changing, and some of the more significant trends are discussed in Chapter 12.

4.2.8 Equipment selection

Selection of equipment must go well beyond considering just insertion rate and capital cost. It must also cover the manufacturing yield, assessing not only the basic cost of assembly but also the cost of any inspection, test and re-work needed to ensure final product quality. Furthermore, selection covering the wider aspects of integration and flexibility is required in order to ensure that investment achieves both today's requirements and also provides an assembly facility which will give the flexibility and scope for tomorrow. The following represent some of these broader issues:

Programming: Is the equipment capable of off-line programming? Are there host facilities available to help the task? Can CAD-produced data be used without modification? Can standard data formats be used?

Data storage and communications: How much on-line memory capacity is there for program storage? Can the equipment link directly to standard communication networks?

Control system: What hardware and software is used and does it conform to standards? Is the system expandable and has the user ready access to it either to modify or add facilities? How does the manufacturer update the facilities provided—indeed, can the facilities be updated?

Operation: Can full remote control be achieved, including status information, and does it adhere to communication standards? What degree of sophistication is built in to allow automatic error recovery?

Accuracy: Has the system sufficient repeatability? Is the absolute accuracy sufficient to remove the need for program debugging?

Component feeding: Can the system be equipped with enough feeders to cater for a large volume single product? Can it hold on-line enough parts to cater for a variety of products? What are set-up times? Can alternative feeders be automatically selected and empty ones replenished on-line? How much do spare feeders cost?

Board handling and fixturing: Can this be automatic? Can conveyors and magazines readily interface to different suppliers' equipment?

Component verification: What is the quality or depth of the verification test? What effect does it have on the insertion rate?

Re-firing: Can the equipment automatically refire? What user controls and reporting are there?

Footprint: How much board area is required for the insertion-head and clinch mechanism?

4.3 Solder and cleaning

Automated soldering falls into two main categories: flow and re-flow. Flow, or wave soldering, applies solder as the PCB passes over a solder wave, whilst re-flow soldering is used to melt and hence re-flow solder that is already present. For SMD applications the solder is usually in the form of a paste, whilst for other components it may be a preform.

4.3.1 Flow soldering

Flow soldering equipment is split into two main categories—single and dual wave. Single wave equipment is applicable for conventional leaded components and for some SMDs. Dual wave systems can be used to solder PCBs with complex SMD packages. In this type of equipment an initial turbulent wave floods solder over the complete underside and a smooth secondary wave cleans off the excess. With both types of equipment some suppliers provide a jet of air, known as an air knife, which cuts through any molten solder that is short-circuiting component pins together. This reduces the incidence of solder short-circuits.

A further development is the agitation of the solder wave. This technique radically improves the quality of joint, eliminating many types of defect.

A further alternative is the solder-cut-solder process. Component leads are left uncut at insertion. At the soldering stage the PCB passes over a bath of solder which floods the underside. Robust rotating cutters trim the underside leads once the solder has solidified. A final smooth and controlled wave provides the correct form of joint, removing excess solder and tinning the bare component leads.

Soldering equipment varies in sophistication. Simple bench-top equipment can satisfy the needs of many users. However, such equipment has the majority of parameters under open-loop manual control, and therefore very careful operating procedures and monitoring are required if reliable defect-free soldering is to be obtained. At the other extreme, equipment is capable of fully automatic operation. All soldering parameters can be down-loaded from the factory computer system. Once the correct parameters for a particular board have been correctly defined they can be automatically set up each time that product is made.

4.3.2 Re-flow soldering

Re-flow methods fall into three types—infra red,[1] vapour phase[5] and laser.[6] Equipment may also be either batch or flow. The first method uses controlled infra-red sources to heat the joints and thus re-flow the solder. Advances in infra-red technology allow precise control of the heating, limiting the stress applied to components.

Vapour phase, as illustrated in Fig. 4.4, utilises a saturated vapour of an inert chemical—the most common being Fluorinert, a fluorinated organic compound. The product to be soldered is pre-heated and lowered into the saturated vapour. This condenses onto the product, giving up its latent heat of vaporisation, and thus rapidly and evenly heating the product. This process can be very closely controlled, as the re-flow temperature, typically 215 °C, is dictated primarily by the chemical composition of the primary liquid. A secondary less dense liquid, usually a chlorofluorocarbon as used for PCB cleaning purposes, provides a vapour seal to retain the primary liquid and thus reduce losses.

In laser re-flow soldering equipment, a controlled laser pulse is applied to each joint to be soldered. This method is considered ideal because the stress applied to components is minimal and reliability is high owing to the joint crystallisation characteristics.

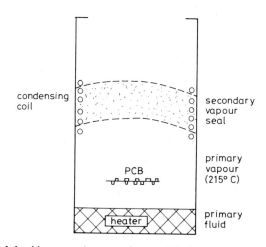

Fig. 4.4 Vapour-phase re-flow soldering

4.3.3 PCB cleaning

Depending on the soldering method and the product specification, there may be a requirement to clean the PCB. A variety of equipments are available in the form of batch and in-line. In-line units can link directly to soldering equipment. The cleaning agent, which may be either water or solvent, can be brushed or sprayed on. Some systems immerse the product in the solvent and may provide ultrasonic agitation in order to enhance the cleaning effect. This can, however, damage some components.

Cleaning often introduces problems from the effect of the water or solvent on certain components. Difficulties can be overcome by careful selection of components or by the use of cleaning systems which only wet the underside of the product.

One of the prime reasons for cleaning is the removal of flux. However, some modern fluxes can be left on the product without causing contact problems during test. In certain circumstances flux must be removed as it can affect the operation of the circuit.

4.3.4 Solder masking

For a variety of reasons some components may have to be fitted after flow soldering. In order to ease subsequent assembly, component holes must be masked. Solder mask, which may be a compound or in tape form, can be applied in a number of ways—by screening, manually or robotically. Removal of the masking material is an additional task which can be carried out automatically by the cleaning process, provided that soluble compounds are used. The aging and solubility of the masking material are important to consider if cleaning is limited to simple spray methods. Other methods such as immersion, brushing or ultrasonics may be required. Ideally masking should be eliminated with careful component specification and product design.

4.3.5 Conformal coating

Some products require a coating to be applied in order to protect against degradation arising from moisture, heat, fungus, grease, corrosive air pollutants and solvents. A variety of equipment is available for this task, the most common being the dip tank. This suffers from the disadvantage of being unable to discriminate between areas to be coated, and therefore spray techniques can be used. Relative motion between the spray head can be achieved manually, by automation or robotically.

4.4 Storage and handling

Storage and handling equipment is different in nature to the equipment types already reviewed, because it adds cost without directly adding to value. However, it cannot be considered in isolation apart from the production equipment itself. An investment in storage and handling must therefore be very carefully considered as part of an overall AMIE implementation plan. Storage and handling equipment should be limited to the minimum consistent with ensuring production flow.

Thus these techniques need applying with care. Automation of storage and handling must be limited to automating the requirement, and not to automating the excess. Furthermore, there is a danger that investment in automated storage and handling can divert capital from value-adding equipment. Requirements and specifications need careful definition and assessment in order to ensure, for example, that store cycle times do not prove to be a limiting factor.

4.4.1 Storage

There are several types of storage—manual, paternoster and automatic. Storage requirements are closely linked to such company characteristics as production volumes and mixes, product physical size, ordering patterns, material and product lead times and production methods.

Manual stores: This is the traditional approach, where material is held on racking and an operator walks to the required location in order to select the required material.

Paternoster stores: These comprise a unit where a fixed series of shelves rotate in order to present material to an operator for picking as required. They are often used to provide storage for small components. They can offer advantages such as stores control, reduced picking time, improved accuracy, security and environmental control. They may be situated remotely from the assembly area or on the shop floor, often being placed adjacent to associated assembly equipment and acting as an on-line integrated store.

Stores such as these could be integrated with the overall computer system. Picking lists, containing such information as bin location, quantity to pick and destination information, would be automatically down-loaded. Operator interaction with the system would be minimal, the main response being limited to confirmation of the required action and possibly some verification input using a bar-code reader.

Automated warehousing: This can provide not only the storage facilities for components, but also for finished product. These stores may utilise high narrow aisle racking with automatic stacker cranes. Pallets are taken as required, and in a dynamically optimised sequence, to either an operator for picking or directly to an automatically guided vehicle for transport to the assembly area.

Compared with manual stores the cost of automated warehousing is high, and therefore the policy regarding stock and particularly WIP levels must be carefully worked out. The benefit of this type of automation will often not be realised unless WIP is reduced.

4.4.2 Handling

Physical movement of product may be achieved in many ways, ranging from manual methods, through conveyorised systems to automatically guided vehicles.

A typical medium-volume configuration may utilise magazines to hold boards for inter-cell transport. Within the cell, boards may be passed between operations by conveyors and mechanisms integral to the equipment itself. Product transported by conveyor may be held in a number of ways—magazines, individual carriers, or conveyor fingers. Alternatively product may be carried on either a wide flat belt or two narrow-edge belts.

Automatically guided vehicles (AGVs) primarily find application in the movement of items between stores and work stations, and in larger installations between work stations themselves. AGVs are increasing in flexibility as the scope and power of communication, control and guidance systems increases. Free-ranging AGVs, which do not follow lines or buried wires, allow a software reconfigurable layout. Radio or other non-contact communication, such as infra-red, permit routing information and status messages to be exchanged.

4.5 Production environment

The quality and nature of production environment is an integral part of AMIE and should not be neglected:

Cleanliness: Contamination and dirt will cause defects. Neither consumption of food or drink nor smoking are considered compatible with quality. There should also be regular maintenance and cleaning of the fabric of the building and light, bright surroundings.

Static: Damage is caused to semiconductor devices by static electricity. The damage can either be so severe as to cause an immediate failure or it may weaken the device, making it susceptible to premature failure. Precautions must be taken to prevent any build up of static, and therefore to remove the risk of damage. Ideally the factory floor should be made of conductive material and all equipment and benching grounded to it. Operators should use earth straps and suitable clothing should be provided.

In order to limit static damage, components and PCBs should not be directly handled. Wherever possible, components should be automatically taken from incoming packaging material and placed directly onto the final location on the PCB.

Storage: Exposure to the air of both PCBs and component leads will cause oxidisation and solderability problems. Raw material should be left in original packing material until required for use, and product soldered immediately after assembly.

4.6 Summary

This chapter has reviewed the major assembly techniques. During planning a balanced view must be taken between the high flexibility and low quality which can result from manual assembly, and the high quality and flexibility of automation.

Automation is often seen as the answer to many business problems. However, in many cases there is a danger that its application will only serve to mask more important issues, such as poor design, excessive inventory, poor flow and low yield. Planning and investment in automation should not be an isolated task under the sole control of production personnel. The opportunities provided by automation can only be maximised if automation is viewed as a part of a company-wide strategy and corresponding understanding and commitment given.

4.7 Bibliography

1. DOW, S. J.: 'The use of radiant infrared in soldering SMDs to PCBs' (Victronics Corporation)
2. HALE, J. E.: 'Complete cleaning of SMD assemblies'. SMT Conference, Long Beach, CA
3. JONES, P. R.: 'Leadless carriers, components increase board density by 6:1', *Electronics*
4. LICARI, J. J.: 'Plastics coatings for electronics' (McGraw-Hill)
5. 'A user's guide to vapor phase soldering with fluorinert electronic liquids' (Commercial Chemicals Division/3M)
6. 'Laser soldering of SMDs' (Carl Baasel Lasertechnik)

Chapter 5

Test

This chapter outlines the various tools available for test, inspection and the management of these tasks. It discusses methods of ensuring product quality by a coherent strategy rather than by using test to eliminate defects.

5.1 Introduction

The traditional role of test and inspection has been to identify and diagnose faults, with elimination by re-work. Diagnosis is increasingly seen as a secondary function, with the prime functions of test and inspection being to certify product conformance and to provide feedback information to control the preceding processes. The emphasis is shifting to enable the process to be corrected as soon as a fault occurs rather than having to correct a considerable volume of product.

In high-volume production environments, the tasks of inspection and test are starting to take place on line, providing a facility for real-time collection and analysis of production performance data.[6] Should faults exceed preset levels, alarms may be given and the process stopped. In lower volume environments the benefit of collecting and analysing production performance information is still of such great importance that the task cannot be neglected, whether carried out manually or automatically. Much modern test and inspection equipment is designed to be integrated to enable the ready collection and analysis of this data.

Computer integration permits paperless repair to be carried out, eliminating much of the time, error and inefficiency normally associated with repair. Board history information can also be obtained as a natural by-product of integration. Definitive performance data produced as a result permits objective decisions to be accurately and rapidly made, thus preventing unnecessary levels of re-work building up.

The advantages of computerised systems over manual systems are very considerable, and can radically reduce the costs and improve the efficiency of repair:

Accuracy: Information captured at source, and as it is created, is intrinsically more accurate. Any manual methods relying on filling in log sheets are very much more open to abuse and carelessness.

Speed: Information transfer is virtually instantaneous. Collection and analysis of data do not have to wait for the end of the week or for the internal mail to operate.

Ease of use: Computerised data-base systems allow ready analysis of information. Analysis tasks which would not even be contemplated in a manual operation can be readily performed. This gives both the vital trend data and the specific information required to focus action on the causes of defects.

In order to certify conformance to specification and also to provide specific performance information, a number of test and inspection techniques can be used. These can range from manual visual inspection through to the use of very high performance automatic test equipment. The main techniques are described below, and Table 5.1 summarises them, ranking them in relation to their inspection and test capability. The choice of methods will be related to individual company characteristics and strategy.[6]

5.2 Test and inspection tools

5.2.1 Goods-inward inspection procedures

Both the direct cost of re-work and the indirect costs associated with production-flow disruption, which result from defective raw

Table 5.1 *Test and inspection application summary*

	Test/inspection equipment type				
	Goods-in	Vision	In-circuit	Functional	Environmental
Component functionality	●	×	●	●	•
Component presence	×	●	●	●	×
Component type	•	●	●	●	×
Manufacturing defects	×	•	●	●	•
Circuit functionality	×	×	●	●	•
Latent defects	●	×	●	●	●

NOTES:
× Inapplicable.
Size of marker indicates relative effectiveness of technique.

materials, force users into considering procedures for ensuring the quality of raw material. Surface-mounted devices, with the associated problems of test fixturing and rework, have only served to increase the need for incoming component quality assurance. For those companies implementing just-in-time manufacturing, component quality assurance is a prerequisite. Quality assurance includes vendor assessment, and this may assume a greater relevance than, for example, incoming component test procedures. Rather than having generic inspection procedures, sampling methods can be employed on the basis of experience.

Goods-incoming inspection procedures are applied for a variety of reasons:

- To verify supplier conformance.
- To identify and eliminate defects.
- To burn-in components, and hence reduce product early-life failures.

Supplier-conformance verification is usually carried out on a sampling basis and may utilise manually fed and controlled equipment. Conformance testing may include monitoring of such physical parameters as component lead solderability and metallisation.

Bulk component testers are usually equipped with automatic handling mechanisms to cater for IC tubes or discrete component bandoliers. Some of these systems have heating chambers to test components at elevated temperature.

Component burn-in equipment is now software configurable to allow virtually any type of logic to be burnt-in using standard hardware, thus eliminating the need for special hardware to be built up in order to test individual devices.

5.2.2 Machine vision

Machine vision is an emerging technique for automatic inspection of physical items.[3] It does this by taking an image of the item being inspected, such as a PCB, analysing the image and comparing the result against a set of pre-defined rules. It is not to be confused with optical-comparator type systems, which aim to compare one image with another. Some of the more advanced comparator systems employ processing functions to cater for manufacturing tolerance and component physical attribute variations.

Many applications of machine vision are possible, but they should be seen in context. Although machine vision can read text, and therefore check for correct IC type, it usually proves to be far more cost effective to control production more tightly, and ensure that only functional devices of the right type are inserted in the correct position. A realistic application of vision would be to inspect for the presence of all underside leads as a final on-line process just prior to soldering. This would give confidence that all required components were present and correctly inserted.

Machine vision is still in its infancy, and to date there are relatively few installations. This will change as costs decrease and programming methods improve. Current methods of programming combine CAD-derived data with expert knowledge obtained from people, with the system building new knowledge by inferencing and user dialogue. Developments in the near future will embrace the expert systems approach, applying artificial intelligence techniques such as heuristic searching.

5.2.3 In-circuit test

In-circuit testers are used to test electrically each individual component on a PCB and also to check that the PCB is free from manufacturing or assembly defects such as electrical short- or open-circuits. A confusion of terms surround this type of equipment, including bed-of-nails testers and manufacturing defects analysers. The term bed-of-nails relates solely to the fixturing method, i.e. the method by which the PCB is physically interfaced to the test system. A manufacturing-defects analyser is similar to an in-circuit tester, but with a lesser test capability. Referring to Fig. 5.1, a manufacturing-defects analyser will probably not be able to effectively test active components or carry out any of the power-on tests; in addition, measurement accuracy for passive components will also be less. Measurement accuracy is also traded off against a significant increase in speed.

The basis of in-circuit testing is to attempt to electrically access and isolate each component on the circuit. Access is through a bed-of-nails to every terminal of the components being tested. For leaded component boards this rarely presents any difficulty; however, access from both sides may be required for SMD-based products. Lack of free board area for test points often makes bed-of-nails fixturing complex and expensive for SMD-based products, thus exerting pressure for alternative test and quality assurance procedures.

The test sequence for in-circuit testers is illustrated in Fig. 5.1. The sequence is designed not only to identify simpler faults first, but also to eliminate the effect of the more likely and basic assembly faults from the testing of complex, and difficult to re-work, integrated circuits. Test starts with power-off testing of basic manufacturing items, followed by passive and active components. Power is applied, and if consumption is within tolerance, testing of active components, such as ICs, proceeds. If a fault is detected there are often diagnostic functions available to rapidly locate the source of the fault.

Individual components need to be electrically isolated in order to test them; this is achieved through the use of one of two techniques, guarding and back-driving. Guarding is used in conjunction with com-

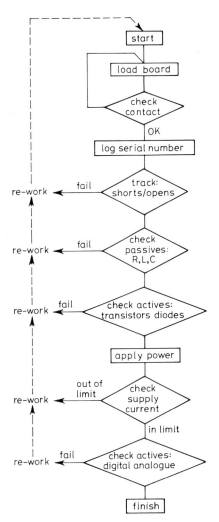

Fig. 5.1 In-circuit test flow chart

ponents where voltage or current are measured to determine the parameter of interest. Node forcing or back-driving is used on digital integrated circuits, where logic states must be forced and measured. Components, such as relays, operational amplifiers and special-purpose devices, such as voltage regulators, fall into a category where neither guarding nor back-driving apply. Individually applicable test methods are required for each component type, and these are often supplied in the test equipment manufacturers' standard libraries. In addition, depending on how the component is used in the circuit, the test may have to be modified each time it occurs on the board.

In-circuit test (ICT) should not be viewed solely as a test function, but also as a means to provide process monitoring. A great deal of specific information about components and the effect of the process is available, since ICT examines every component on every board. This should be used as feedback data for component specification and procurement, and to improve the assembly process.

A major advantage of in-circuit testers lies in the ease of generating test programs. Many systems have extensive libraries of device models and provide automatic test program generation (ATPG) facilities direct from CAD-produced data. This will often include fixture building data such as NC drill and wire-wap NC programs.

5.2.4 Functional test

Functional testing is concerned with verifying that the unit under test (UUT) has the correct transfer function; namely that, for a given set of input stimuli, the correct output response will be observed. This technique is equally applicable to both analogue and digital PCBs; however, the particular test-system hardware configuration will affect the scope of the test capability. If the UUT is identified as faulty, the ATE can be used to identify or diagnose the cause of the failure.

Fault diagnosis can either be performed by skilled test engineers referencing the test program and circuit diagram in relation to the UUT operation, or by the use of diagnostics resulting from fault simulation. This, linked to rule-based systems and heuristic algorithms, enables an operator to rapidly probe back down a circuit chain until the fault is located.

For a complex PCB, using state-of-the-art ICs, writing an effective functional test program may take, say, ten times as long as for an in-circuit test program. In order to cut this time, and hence the product introduction time cycle, data captured and created by the designer during the initial stages of circuit design and simulation can be used as the basis of the test program. A growing number of vendors provide software to convert simulation files into functional test programs; however, some further programming effort is still usually required in order to fine-tune the test program in order to take into account parameters such as timing, speed and skew. Unless these factors are taken into account during the design phase, use of the simulation data may be difficult if not impossible.

5.2.5 Combinational test

Whilst in-circuit and functional test have in the past been individual pieces of equipment, a growing number of combinational systems are available which combine into a single machine the capability of both techniques. A new generation of fixtures is available which permit dual stage access to the PCB. This is required because some of the test pins required for in-circuit purposes will unduly load the circuit, thus adversely affecting the operation of the circuit. Combinational testers can offer a substantial advantage in reducing the amount of board handling and achieving a higher utilisation of equipment. Balanced against this is the fact that, during the in-circuit test phase, the investment in functional test capability will be idle and vice versa.

5.2.6 Environmental stress screening

Faults fall into two categories: those which are active and continuously visible and those which are intermittent or latent in nature. Test equipment of the types already mentioned will all identify the active faults. However, if the latent or intermittent faults are to be identified and eliminated, other techniques are necessary. Latent and intermittent defects result from several causes, and become apparent as the normal operational cycle of heating and cooling puts stress on material, causing eventual failure. Typical latent and intermittent defects include poor solder joints, defective IC wire bonds, semiconductor impurities, semiconductor defects and component drift.

The outcome of many studies on life expectancy and failure patterns of electonic items has resulted in the Weibull or bathtub curve, shown in Fig. 5.2. Failures fall into three categories. The infant mortality phase accounts for the greatest number of failures, while the failure rate falls to a low and constant level for the mature phase. Finally the rate starts to rise as the product begins to wear out. The aim of environmental stress screening is to ensure that all the infant mortality failures will manifest themselves before the product is shipped, but not to over-stress to the extent that good parts will be damaged or their lifetime shortened.

The benefit of stress screening extends far beyond reducing site failures and the associated cost of repair or replacement. It can serve to reduce product development time cycles by simulating extended operation using the technique of time compression.

Of the many techniques that exist, random vibration, high temperature, temperature cycling, and electrical stress appear to be the

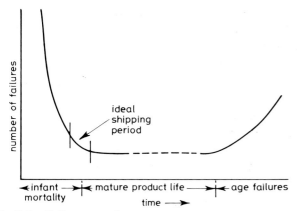

Fig. 5.2 *Failures vs. time*

four most effective techniques—these can be used in isolation, or may be combined.

Random vibration precipitates mechanical latent defects such as cold solder joints and bond wire problems. It is usually limited to aerospace and military applications.

The most traditional form of stress screening is burn-in or heat soak, in which the operating temperature of the board is raised to a pre-determined level for a specified time, usually 48–72 hours. Burn-in is generally a good trade-off between reducing the reliability of good components and precipitating a fair number of latent faults. However, mechanical defects such as cold solder joints or bond wire problems are not accelerated by burn-in, since only limited mechanical expansion and contraction occurs.

The most successful form of stress screening is temperature cycling or thermal shock, because both contamination and mechanical faults are precipitated in a relatively short time. Unlike burn-in, where the static temperature level over a period of time is the important factor, here the rate of temperature change is critical. In addition, low temperature plays an important role. Most temperature-cycling systems range from −20 °C to +125 °C. Some go as low as −40 or −50 °C. Because greater stress is caused by changes in temperature than by static conditions, temperature cycling identifies a wider range of problems than does burn-in.

Electrical stress consists of two aspects: power cycling and dynamic exercising. Power cycling turns power on and off repeatedly while the temperature-cycling progress is taking place. Dynamic exercising involves sending electrical stimuli to the inputs of the board under test. Signals which cause devices and junctions to change state create stresses within device junctions not otherwise affected by power or ambient temperature. Like power cycling, dynamic exercising augments the effects of temperature cycling.

5.3 Re-work

Computerised tools are available to aid the task of re-work. Diagnostic data produced by the ATE can be assessed in several ways. Traditionally paper was involved—either handwritten or automatically printed. Today's paperless systems take the fault data associated with an individual board and pass it to a local data base. The re-work system, which can range from a simple VDU screen through graphical displays to a guided-light and parts-dispensing system, enables an operator rapidly to identify the item requiring attention. For example, a graphical system will display the track or component layout and highlight the fault. The operator will repair the fault and input, to the system, data concerning the cause of defect—solder short-circuit, incorrect component etc. Operator input is usually by means of a bar-code reader, thus speeding the task and reducing the incidence of error.

These systems offer considerable advantages not only in reducing the time involved in physically locating faults once the ATE has electrically identified them, but more importantly in rapidly and accurately capturing specific information concerning the nature of the fault, allowing effective real-time control to be effected.

5.4 Test area data bases

The test and re-work area produces a considerable volume of information concerning the nature and cause of failure. This information may be treated in one of three main ways:

- Discarded.
- Corrupted.
- Used effectively.

In the manual and non-integrated environment, most of the information produced is either discarded by not being recorded or is corrupted—the latter because repair record sheets are not filled in until usually well after the event, by which time the frailty of human memory will have had an effect. Effective use can only be made of the data if it is captured once at source, as it is created, and fed into a computer data base for access and analysis.

Either standard data-base systems or proprietary test area management systems allow the raw data to be converted into vital and focused information, thus giving the background to enable problems to be rapidly identified and dealt with. Some systems allow real-time alarms to be set to give immediate indication of gross manufacturing problems.

Reports are available which analyse the data base in a variety of ways, in order to identify problems before they get out of hand. The reports can point to the source of problems, enabling action to be taken to correct, modify and enhance, thus reducing costs and improving product quality. Reports which are designed to focus on particular aspects of the production process allow monitoring of productivity, yield, failure data, test times, repair times etc. These can be formatted to suit the particular requirements of the application.

For example, a summary report could show whether a problem originated in a first or subsequent test pass. A first-pass problem would generally stem from a manufacturing defect, while a subsequent-pass problem can usually be traced to testing, repair or design problems. Other analyses of the data base could be by component failure or cause. These would be analysed further to consider failures across a single PCB type, a family or the complete range of product. Information can be obtained or failures analysed down to an individual assembly work station.

This kind of vital information, not readily available without a computerised data base, allows the allocation of resources in the most cost-effective and timely manner.

5.5 Strategy

So far we have concentrated solely on the individual tools rather than on how they might fit into the overall environment—the test strategy. Strategy is not primarily concerned with the question of 'how to test' but is aimed at ensuring the quality of design and production. For example, raising the production yield from a current average of, say, 60—90% will have major impact on the viability of the whole operation. This 75% cut in defect rate will reduce the time spent on diagnostic test by a similar amount, and can have a dramatic effect on equipment utilisation. The disruptive effect of re-work will be greatly reduced. The impact of this on production as a whole will be considerable. The greatly reduced disruptive effect of diagnosis and re-work will ease work flow, reduce bottlenecks and permit the passage of work through test to be effectively planned and controlled.

Raising the yield to an acceptable level will require action to be taken to identify and attack the root causes. This is illustrated in Fig. 5.3, with key areas outlined below:

- Quality of design, with manufacturing requirements built in and depth of engineering analysis to enable 'right first time'.
- Low defect rates of incoming components by supplier interaction and quality assurance procedures.
- Good assembly practice—controlled storage and handling, preventive maintenance of equipment, process monitoring, minimum process time.
- In-line inspection for real-time process monitoring, with re-work at each process step.
- Management concentrating on prevention rather than test, diagnosis and re-work.

An important tool in bringing quality into control is statistical quality control (SQC). In this technique, the nature and cause of defects are analysed in order to provide a statistical basis for action. Depending on the application, either manual recording or computerised methods may be employed, with the latter offering increased accuracy, ease of use, and flexibility in data analysis.

For example, one company achieved a 3:1 reduction in solder defects by basic production engineering, and followed this by an SQC program to further reduce the defect level. This resulted in a 50:1 reduction, with the defect rate finally standing at 100 parts in 10^6. This low defect rate enabled the additional, unreliable and subjective process of joint touch-up to be eliminated. In addition, the implementation of an SQC program in the auto-insertion area reduced insertion defects from 30 000 to 5600 parts in 10^6.

Based on a strategy for high yield, the exact test route for a particular product can be developed with the objective of rapidly and cost effectively diagnosing both active and latent faults. There will be no one optimal test

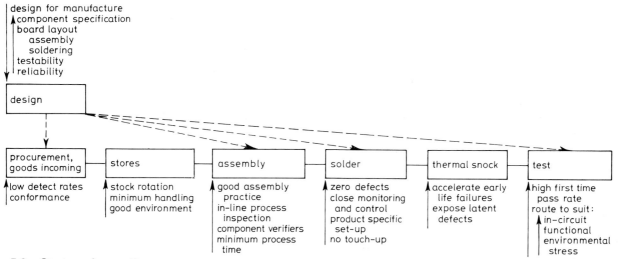

Fig. 5.3 *Strategy for quality*

route, but routes will depend not only on the specific nature of each manufacturing environment, but also on the products themselves. It is not unusual to find three or four substantially different routes, depending on the product type, manufacturing route, electrical characteristics, complexity and market.

5.6 Summary

Test is no longer an isolated process which occurs after production has finished, but is a vital integral portion of production. The role of test should be viewed in the context of ensuring quality rather than identifying defects. Quality is the key, ranging from design, through procurement and into manufacture; then, and only then, is the question asked 'how to test?'

5.7 Bibliography

1. CONSOLLA, W. M.:
2. DANNER, F. G.: 'An objective PCB testability design guide and rating system'. Rome US Air Development Centre Report IR-79-327
3. DENKER, S. P.: 'Automatic visual inspection of assemblies containing SMDs' (Cognex Corporation)
4. FULLER, T.: 'Correct the process, not the product', *Electron. Production*, Sept. 1984
5. MANGIN, C. H.: 'Implementing high yield electronics assembly at low cost' (Ceeris International)
6. PYM, C.: 'Strategies for electronics test' (McGraw-Hill)
7. VANDERMARK: 'The removal and reattachment of SMDs' (Nu-Concept Computer Systems)

Chapter 6

Manufacturing resource planning

This chapter discusses the problems, common to many companies, of excessive cost and time. The tools which are available for production planning and control, which seek to overcome these problems, are outlined.

6.1 Introduction

The five preceding chapters have reviewed some of the most relevant techniques encompassed by the term AMIE. Techniques such as these cannot in themselves be used to achieve efficient and on-time production without some sort of pre-planning. This is where manufacturing resource planning (MRPII) fits in, by enabling more effective use to be made of such resources as working capital, equipment and people.

For over 25 years, computerised systems have been available which addressed the basic tasks of ensuring that sufficient material was on hand to meet the product-demand requirements. This technique is known as material requirement planning (MRP). MRPII evolved from this through a series of steps. These included the ability to not only plan and schedule for material, but also for production capacity. Feedback functions were provided to help ensure the constant validity of plans and data. Financial control gave a translation of the physical production plan into monetary terms. The most recent significant development lies in the ability to simulate and ask 'what if' questions. This ability provides the accurate background data necessary to make effective management decisions. Fig. 6.1 reflects the prime operation of MRPII, ranging from initial business planning down to asking the 'what if'? questions. These are concerned with how best to utilise material, equipment and people resources in order to meet the required customer demand and minimise cost and risk. It is a closed-loop heuristic process,

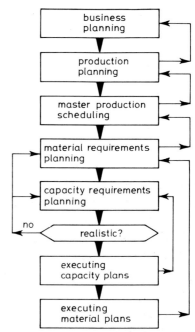

Fig. 6.1 MRPII

learning from previous experience in order to provide a more accurate statistical data base upon which to make future judgments.

MRPII is an integrated system, spanning the complete business. The major functions of it are shown in Fig. 6.2. At heart of the system is information relating to the company, products, suppliers, customers and the marketplace.

The benefits to be gained from MRPII are some of the most significant to be obtained from all the techniques mentioned in this book. Properly implemented, MRPII will provide a payback period usually measured in months. Much of the benefit comes from inventory savings. However, other far-reaching advantages are to be gained, such as achieving due dates, which, though more difficult to cost or quantify, can provide a greater degree of competitive impact.

In common with other systems, many MRPII applications have failed to live up to

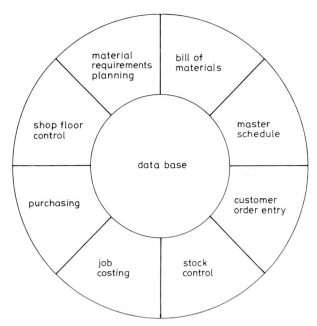

Fig. 6.2 *MRPII basic functions*

their expectations. In order to help assess the performance of an MRPII system, the ABCD classification of system performance has been developed.[2] This provides a reasonably objective method of self assessment. The classes are broadly defined as follows:

Class A: The system is used by all functional areas of the business, i.e., sales, finance, manufacturing, purchasing and engineering. It provides the plan against which activities take place. The system provides 'what if?' capability in order to provide answers to questions related to business planning.

Class B: a class B implementation will have many of the modules in place. It differs from class A because management does not use the system to run the business directly. The system is probably viewed as a production and inventory control system. Additionally, although some production scheduling will be performed, the shortage list is usually used as the production schedule. The substantial benefits of improved productivity and reduced inventory are rarely fully realised.

Class C: a class C company will use MRP primarily as a purchasing control system rather than for any type of scheduling. They will have no closed-loop control of production.

Class D: The system is reserved for the data-processing function. Records are usually poor, and any schedules are wildly inaccurate and are not used. However, many class D users have very successfully achieved re-implementation with basically the same hardware and software, but by properly carrying out the tasks of education, training and accurate data-record creation. Failure has invariably resulted from a lack of commitment, planning or implementation skill.

6.2 Business problems

The multiplicity of planning, procedural and operational problems found in most companies merge to give the two general problems of excessive cost and time. These result in a reduced profit margin and market share as customers find more responsive suppliers offering better value. Excessive cost and time stem from a number of factors, often interrelated and showing a company out of control:

Stock control: Despite excessive inventory, shortages are accepted as the norm. Constant inventory reduction programmes never achieve the required result, with stock always showing a sizeable proportion of obsolete or slow usage parts. Excessive manpower is spent on stocktaking, perpetual inventory analysis and the like, with discrepancies always being uncovered, but no net improvements made.
Deliveries: Only a small proportion are on time despite much overtime. Shipping forecasts are rarely fulfilled. Many achieved schedules are for stock items not required to meet current shortages.
WIP level is excessive, resulting in jamming of facilities, masking of true problems and impossible tasks for progress chasers.
Productivity is low because of material shortages, poor planning and lack of control.
Production peaks during the last week of each month, tying up valuable management effort in fire-fighting.
Paper work is excessive, out of date and inaccurate.
Data: Statistics and information across different departments are inconsistent, out of date and inaccurate.
Change control and management is ineffective often resulting in excessive re-work and stock write-off.

Such a list is far from exhaustive and reflects only a proportion of the problems arising from a lack of planning and control. However, through the application of proven and practical techniques for planning and control, such problems can be reduced or eliminated.

6.3 MRPII: Practical benefits

The major initial benefit resulting from MRPII is usually the capital released from inventory and WIP reductions. However, significant benefits can be seen throughout the company. Management can now focus their activities on planning and profit-making activities instead of depriving the business of its key resources by constantly trying to plug holes.

The benefits of MRPII can be substantial, and the following are typical of what should be expected.

Stock is now under control and savings of at least 20% are usually achieved.
Deliveries are now on time, improving customer image. Partial shipments are reduced with a corresponding reduction in paperwork, staff and freight charges.
WIP levels are down, making problems visible, improving quality and freeing storage space for production.
Productivity of staff, both direct and indirect, and of plant is increased.
Production is now smoother, with a reduced need for overtime.
Paper work is reduced, with fewer exceptions to deal with and less batch splitting. On-line terminals are used in place of out-of-date, costly and voluminous paperwork.
Data is now accurate, up to date and consistent across the company.
Change control is easier. Management knows where the product is, and the window for physical change decreases very considerably.
What if facilities now exist to allow management to fully evaluate the effect of alternative plans.

6.4 MRPII: Functional description

MRPII, as Fig. 6.2 shows, consists of a number of functions or modules, which are integrated with a single data base in order that they can function as a single cohesive system.[1,3] The following are typical of the more significant functions.

6.4.1 Bill of materials
Bill of materials (BOM) is primarily a parts list, but can often be extended to include information such as tooling, test, manufacturing or packaging requirements. The manual data entry methods associated with MRP have been eased with the provision of part structure duplication and modification. Furthermore, integration with CAD eliminates most of this data entry effort. 'Where used' information is invariably provided, thus easing tasks concerned with component changes or obsolescence.

6.4.2 Material requirements planning
Material requirements planning seeks to simulate the manufacturing process in order to formulate the minimum supply levels required to permit the lowest inventory level. Associated with each item will be lead times for administration, processing and stocking. Complex assemblies cater for individual sub-orders lead times. This module may be required to be driven from a variety of sources. Typically, the master production schedule, customer orders, sales forecasts or work orders from the shop floor can all be used, although ideally demand should be stimulated by customer orders.

6.4.3. Master schedule
The master schedule takes into account the production demands and the constraints of individual activities within the company, in order to produce plans which make efficient use of resources and meet long term corporate objectives.

The scheduler should consider such key factors as working capital, production capacity and customer demand, in order to build a model which will clearly show the range of alternatives and the effects of each. The schedule should show both the day-to-day requirements and reflect long-term forecasts.

Capacity planning is a key feature for the scheduler; production is simulated according to the master schedule and critical work-centre bottlenecks are identified and analysed. This gives the opportunity to re-schedule, or, in the case of recurring problems, it can act as the stimulus for long term corrective measures.

6.4.4 Shop-floor control
This module allows work on the shop floor to be tracked, thus providing feedback data for input to the scheduler. Additionally, it can provide valuable management information, showing performance trends. Plants operating a JIT system will usually restrict tracking points to two: material issue and job complete. Even a factory manufacturing many hundreds of thousands of PCBs per annum will find little

need for detailed work tracking if the floor-to-floor time is commensurate with the in-process time.

6.4.5 Customer order entry
This is concerned with the entry, co-ordination and tracking of orders from receipt through to dispatch of product. Order-receipt routines should enable rapid entry of order details with prompts to ensure that all the required information is entered. The system should detect the majority of order input errors. More sophisticated systems integrate order entry with other modules to enable 'what if?' type questions, and thus provide rapid and accurate customer response.

6.4.6 Purchasing
Purchasing, interfaced with the material requirements planning module, will generate order documents and provide facilities for entering material receipts into stock. Usually included in this module would be vendor performance analysis concerning such parameters as quality, performance and delivery achievements. The system should provide management information and exception reporting to initiate expediting, for example.

6.4.7 Stock control
Stock control seeks to improve efficiency by ensuring stock levels are commensurate with the master production schedule. It will aim to minimise overstocking—thus saving capital unnecessarily being committed as WIP—and understocking—thus reducing material shortages, the failure to achieve due dates and the disruptive impact of shortages on production flow.

Specialised options may be required to control, for example, perishable items or material purchased under contract, and so as to facilitate traceability.

6.4.8 Job costing
This module calculates costs, either from bills of materials and routings or by accumulating information during manufacture, taking into account material, labour and overheads. Costs can be standard, actual or average.

It should be possible to monitor costs during processing and provide reports on a periodic basis, a job basis or by variance. The latter would typically provide indicators on material, labour and overhead efficiencies.

6.5 Systems integration

There is considerable benefit to be gained from integrating an MPRII system with functions that are outside the direct manufacturing area. This will enable common information to be shared and data to be transferred between systems, and thus eliminate the problems of delay, error and discrepancies associated with manual data transfer and fragmented, isolated system. Fig. 6.3 reflects some of the more significant links that are often established. Some of these will be inter-computer, whilst others would be to a terminal or printer in another department.

Fig. 6.3 *MRPII: integration*

Integration of design allows the automatic and rapid creation of bills of materials and routings from CAD-generated designs, whilst access to the MRPII parts data base permits design to use accurate and up-to-date part-related data such as cost, preferred types, usage trends etc.

Management and financial accounting are facilitated by the integration of the MRPII with accounting systems. This will permit automatic updating of ledgers, issuing of invoices and payments, and keeping track of debtors and creditors. Management information in the form of budgetary forecasts, planning models and cost-related performance indicators can all be provided.

6.6 MPRII: Summary

The potential benefits of MRPII are very significant, often resulting in at least a 2:1 inventory reduction and an improved responsiveness to customer demand. However, benefits such as these can never be achieved unless the system is implemented correctly. Top-level executive backing and commitment of the necessary resources are essential. Furthermore, the system will not in itself resolve bad management practices, nor will it succeed if it is unnecessarily complex or rigid. Successful implementations are many, and provide those companies with a significant competitive advantage.

6.7 Bibliography

1. CORKE, D. K.: 'A guide to CAPM'; (I.Prod.E., ISBN 085510 028 1)
2. WALLACE, T. F.: 'MRPII: Making it happen' (Oliver Wight Publications, ISBN 0-939246-4-X)
3. WALLINGTON, D.: 'CAPM in practice'. Proceedings of EPEE, 1985
4. WALLINGTON, D.: 'Winning ways with CAPM' (Inbucon Management Consultants)
5. 'Justifying MRPII' (Manufacturing Sciences)

Chapter 7

Just-in-time manufacture

Technology never forms a total solution. This chapter discusses just-in-time manufacture, a concept which can radically and positively affect the operation and viability of a company.

7.1 Introduction

Implementation of an advanced manufacturing approach will never wholly provide a cost-effective customer response. Indeed, without careful planning and organisation, much automation will only serve to limit flexibility owing to longer set-up times. This can in turn inhibit the manufacture of small batches—a key element in many companies' ability to provide rapid response. There is often a pressure to keep high-speed automatic assembly equipment running, apparently to justify its presence even though it may be manufacturing goods speculatively for stock, or only partially assembling products for which some components are still not available. The first procedure builds up unnecessary WIP, whilst the latter suffers from additional assembly cost in hand assembly of the components as they arrive, and from poor soldering quality as product oxidises while it waits.

There are many case studies of companies which have tried, but failed, to resolve the problems by applying automation without tackling basic problems. However, many case studies illustrate those companies which have both applied the technology and also resolved the basic issues. They indicate considerable savings, as shown in Table 7.1, often for little investment. A £100 million turnover company may well benefit annually by £6 million with a one-off inventory saving of £4 million. Set against this is an implementation cost of some £3 million spread over five years. In another instance, a medium-sized manufacturing unit invested £150 000, which was paid back in less than six months with the following impressive results:

Space saving: 14 000 ft²
Inventory reduction: £2 million
Labour reduction: 10–20%
Quality improvement: 60%

Most case studies indicate that, even from pilot projects, an average payback of less than nine months can be expected.

The catalyst for today's approach to manufacturing philosophy was the Japanese push towards quality during the 1960s, which extended in the 1970s to reducing inventory and manufacturing costs.[14] Three common interpretations of this philosophy relate to flow, quality and waste:

Flow: Treat the whole of the business as a flow process rather than batch oriented.
Quality: Implement total quality control so that the customer, internal or external, always receives 100% quality.
Waste: Eliminate waste, by using the absolute minimum amount of equipment, labour, material, space and time to add value to a product.

The implementation of these philosophies comes under the umbrella term of just-in-time (JIT). In many environments there is a dra-

Table 7.1 *Typical benefits of a JIT implementation*

Inventory	down 90%
Cost of sales	down 15–40%
Production lead time	down 90%
Employee productivity	up 10–30%
Set-up times	down 75%
Space utilisation	up 50%
Purchase-price reductions	5–10%
Quality	up 75–90%

matic difference between the sum of the individual process times and the floor-to-floor time. For example, it is not uncommon to find a ratio of 40 or 50 in an electronics manufacturing plant. JIT seeks to reduce the differential in order to bring the floor-to-floor time closer to the sum of the process times.

Just-in-time brings, in addition to the direct benefits already mentioned, very significant benefits in flexibility. Components need not be allocated until required by customer demand, and therefore optimum utilisation of stock is possible. Reduced process times also cut down the window in which modifications can be introduced, and reduce the usually onerous task of managing change.

Implementing just-in-time is not limited to the production floor alone. Where is the advantage of, for example, being able to manufacture in five days if a customer order takes ten days to pass through the administration system? JIT should be applied as much to the office as to the factory.

JIT is too important an opportunity to miss. Table 7.1 illustrates typical benefits, such as inventory and lead times being cut by 90%. Without JIT the adoption of some of the approaches described in this book have little value. Furthermore, JIT is not restricted in application to mass production, but has widespread application for batch production. This chapter reviews the main JIT techniques, and then goes on to consider the implications of JIT for computer systems.

7.2 Implementation

Implementation of just-in-time should affect all aspects of a company's operation:[9,13]

- Top management
- People development
- Total quality
- Waste elimination
- Manufacturing route simplification
- Problem visibility
- Continuous improvement
- Supplier relationship

7.2.1 Top management
JIT can only succeed when it is initiated from the top down. Management needs to have very clearly defined objectives of what is to be achieved and the methods by which the achievements will be made, and needs to be convinced of the need to improve and achieve the objectives.

Enthusiastic commitment from the top, with a constant monitoring of progress, evaluation of results and re-assessment of methods will help to engender commitment at lower levels —a key element in the success of the project.

7.2.2 People development
Whilst JIT is managed and initiated from the top down, it has repercussions throughout the company. Fundamental to the success of JIT is employee involvement and co-operation. 'us and them' attitudes must go, and a corporate approach stressing the importance of team work with the goal of customer satisfaction should become priority.

Effective training, not only in job-specific techniques but also in quality improvement and problems solving, is important. Education is necessary in productivity and cost issues, the use of data-analysis tools, and in understanding and utilising the information that is produced.

Such incentive schemes reward by volume produced, i.e. piecework, are invariably stumbling blocks. Payments should be centred on worker flexibility, and quality of output geared to meeting company profit objectives. Some form of quality circle can usefully be employed to increase and formalise employee involvement. Involvement and commitment at all levels is helped by creating an environment in which all employees are aware of the importance of the contribution they make to the success of their company. Sensitive and firm resolution of demarcation issues, coupled with work-force flexibility, can create a very responsive environment. When a problem arises it can be dealt with immediately by people on hand, instead of having to stand idly by while the correct grade of person is found. In Japan the separate cadres of blue- and white-collar employees are being replaced by single-status arrangements.

7.2.3 Total quality
Just-in-time relies on continuous flow concepts. Inconsistent quality, whether it be in raw material or in the process product itself, will disrupt flow and stop JIT from working at its most effective level.

Total quality is the main key to introducing JIT, with the aim of identifying and removing the source of failure rather than correcting the product. Information forms the analytic basis for identifying the source of defects. Computerised systems, as described in Chapter 5, are invaluable and can be applied widely. Test, inspection and performance data are collected, collated and analysed to produce useful information, and identify trends so as to correlate failures with process variables. This information can be presented in a variety of ways to aid the rapid diagnosis of faults.

Inspection, as a separate operation, wastes not only resources but also reduces flexibility. Self, or in-line, inspection can identify problems at source. This enables remedial action to be taken before a large amount of defective product has built up, thus eliminating unnecessary re-work and scrap. In the event of a serious fault being identified, the flow must stop until the fault is rectified. Putting the onus on the operator for quality has important spin-off benefits, as operators become more involved in the process, and are able to diagnose the basic problem and recommend solutions in advance of a production engineering investigation.

The concept of quality must not be limited to what the end customer receives, but must extend to wherever material, or information, is passed from one process or person to the next. The view of 'who is my customer?' must be broadened to include internal as well as external customers, with employees more aware of their responsibility to produce quality items right first time.[14]

Bad designs may automatically induce defects. Quality must be built in from the conception of a design. A high class of workmanship is required, using, where appropriate, computer-based tools. Use of CAD can eliminate engineering changes, with their heavy burden of labour and severe production disruption. Production and design need to communicate constantly. The use of CAD, with a comprehensive engineering data base reflecting manufacturing requirements, will help to ensure that quality is built in and that optimum process flexibility is maintained. As will be discussed in detail later, supplier reduction and rationalisation can help quality assurance and enhance quality.

Finally, flow will never occur if the plant is unreliable. Planning and execution of preventive maintenance are as critical as material and production planning.

7.2.4 Waste elimination

One of the major aims of JIT is to eliminate waste. Waste may be defined as any material, action or process that does not add value to the manufactured product. Waste can be eliminated or substantially reduced in most functions, because:

- Long and complex material flows waste labour, space and capital
- Excessive inventory wastes assets, space and flexibility
- Extended times waste flexibility and quality
- Careless workmanship wastes time and material
- Defective processes and material waste raw material, labour and resources
- Modifications waste material, labour and resources
- In the context of adding value, paper work is waste and should be minimised
- Set-up time wastes production resources
- Unwarranted variety may waste resources
- Separate inspection stages waste resources and lose flexibility
- Lack of value engineering wastes resources and material.

In analysing waste, it is important to differentiate between cost and value. In the manufacturing environment cost is added by receiving, inspection, queues, handling etc., whilst value is added by assembly operations, soldering etc. Waste reduction is aimed at minimising cost whilst maximising the added value, and should not be limited to the manufacturing area.

One of the major sources of waste in the electronic environment arises from engineering changes. It is not uncommon to find 50% of engineering time being associated with change implementation, with a considerable amount of effort required by configuration control, inspection and production management to ensure that the finished product is at the required status. A key area of waste elimination will come in improving the quality of workmanship during the design phase, aiming to fully prove the design before production.

7.2.5. Manufacturing-route simplification

Reducing lead times and cutting inventory levels depends on simplifying the manufacturing route. Simplification can involve drastic changes in traditional methods and approaches which have hidden lack of productivity and inefficiency.[13] Process simplification is based upon several simple and common-sense precepts:

- Simplify flow and layout
- Reduce material movement
- Eliminate queues
- Minimise set-up
- Ensure flexibility of operators
- Design for manufacture
- Use of new technology

Intimately related to route simplification is the philosophy of how flow is created. Is work pulled through on a customer demand basis—'just-in-time'—or is it pushed through on the basis of often irrelevant manufacturing schedules based on inaccurate sales forecasts—'just-in-case'? The pull environment encourages flow because the effect of any disruption is immediately apparent, thus causing attention to be rapidly focused on eliminating the disruption. In the push environment anything except a gross problem is masked by excessive inter-process WIP buffers.

The Japanese Kanban principle has widespread application in pull-type environments. For example, material supply may be on a two-bin basis—whenever an operator empties a material supply bin it is placed at a refill point for the materials handler to collect and refill, whilst the operator uses material in the second bin. In a multi-stage production process the Kanban signal to make another item comes from a down-stream demand. The Kanban, or pull signals, can take many forms. For example, in addition to the two-bin system described above, it could be by electronic data communication or by printed Kanban cards.

Simplifying flow and layout is considerably eased by the adoption of Kanban and pull principles. Batch and JIT material flow are illustrated in Figs. 7.1 and 7.2, respectively. Gone are queues, incomplete kits, line stocks and solder touch up, to be replaced by fast efficient demand-oriented production.

Such a system can run in a make-to-order process, bringing batch sizes down to very low levels and in many cases to one. This can create problems with some of the high-speed auto-insertion equipment, generally thought of as being inflexible and dedicated. Steps can be taken to overcome these drawbacks. For example, one major manufacturer auto-inserts well in excess of 90% of all components on batch sizes of down to one, having first:

- Dedicated parts feeders for preferred components
- Created design control and incentives to encourage the use of preferred components
- Linked CAD equipment to the insertion equipment in order to provide all assembly programs, thus eliminating any manual programming or etching.

Processes for manual operations need to be established with flexibility in mind. The preferred method is the U-shaped line, where

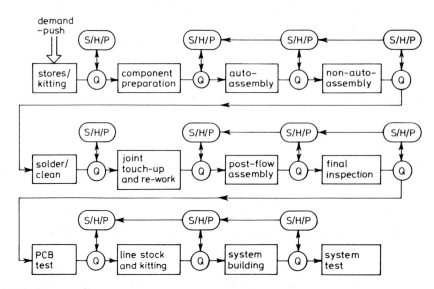

Fig. 7.1 *Batch process route*
S/H/P: Shortage, hold and problem buffer areas
Q: Normal queue

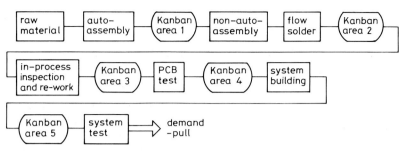

Fig. 7.2 *JIT production flow*

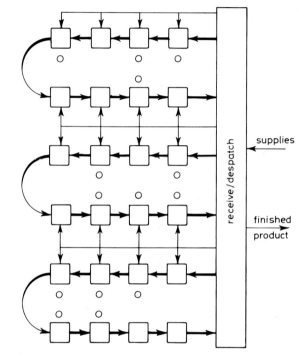

Fig. 7.3 *Group-technology U lines*
← Material flow
→ Product flow
○ Operator
□ Work station

output is varied by adding or subtracting operators. Operators can move from one work station to another, as shown in Fig. 7.3. A team approach is readily developed as imbalances and quality problems quickly become visible. The layout helps to promote self-detection and correction of defects by the operators, who are kept more alert by the nature of a more varied task.

The tradition approach to production, based on producing kits for large batches, is no longer relevant in a JIT small-batch environment. Steps must therefore be taken to introduce, at a physical level, line stocks and, at a design level, a realistic and comprehensive component-rationalisation programme.

Manufacturing route simplification may involve taking very drastic action over material-storage and handling systems. Investment in these areas has on occasions only served to automate the waste between the profit-making processes, and in so doing has diverted capital from improving the process into masking the problems. Automated storage systems, conveyors, racks and WIP containers all have their place, but usually only to a very limited extent in JIT implementation. They can be a real hindrance to improvement, process linking and team building, whilst they consume support staff and overheads. As fast as many companies are sinking capital into automating the waste, the world-class manufacturers are removing unnecessary non-profit-making storage and handling overheads.

The maximum benefits possible from simplifying route and flow will never be achieved without a very intimate relationship between design and manufacture. Design must be based on a very thorough understanding of the capability and limitations of the manufacturing process. It must consider not only what the product will look like and how it will function, but how it is to be assembled. This implies a re-deployment of resources—away from sorting problems on the shop floor and towards building knowledge of the manufacturing process into the engineering data base and new designs. A significant difference in the deployment of engineering resources is seen between various nations. In Japan 90% of the industrial engineers' time is spent trying to reduce set-up time; in the USA, however, 75% of all cost-reduction effort goes into reducing labour costs, even though these only account for 10% of the average company's costs (see Fig. 7.4).

Design for manufacture covers more than direct assembly tasks. It also tackles other issues, such as reducing set-up requirements and the number of process steps. Standardisation of product and components is needed wherever possible. Basic questions need to be asked—are ten different types of the same value of resistor really needed or could a single, and possibly more expensive, type

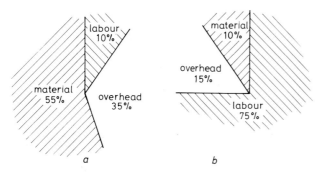

Fig. 7.4 Cost allocation
 a Costs
 b Cost-reduction effort

possibly be cheaper when the total process is costed? Very rapid component development makes rationalisation difficult, but certainly not impossible. The greater the component variety the more complex is the task of design, procurement, management, planning and manufacture.

7.2.6 Visibility of problems

Problems are masked by excessive inventory, because, instead of a problem being recognised and dealt with, other production takes over and the problem is pushed aside. Minimum levels of WIP ensure that, when a problem arises, it is dealt with immediately whilst it is fresh in people's minds, and that maximum attention is focused on it. A problem-solving attitude is fostered in teams and traditional demarcations are broken down.

Effective use can be made of information produced by making that information visible to all. Histograms, Pareto and pie charts indicating performance can usefully be displayed on the production lines, with operators being trained in understanding the significance and use of such information.

7.2.7 Continuous improvement

Just-in-time has been described not as a goal but as a continuous journey. Although goals and objectives can be set it will be found that further improvements are always achievable. The direct involvement of all concerned in the task of manufacture is important. Input from operators themselves concerning improvement to products and processes will form a vital part of a JIT implementation.

7.2.8 Supplier relationships

The steps described so far have been involved almost entirely with internal relationships and organisations. The last, and probably the hardest, to overcome are the traditionally suspicious attitudes between supplier and customer. These are based on the supplier endeavouring to supply the least for most and the customer obtaining the most for least. Such attitudes hinder the stability and quality of relationships, with the result of inhibiting JIT because of delays and poor quality. With raw material costs accounting for a major proportion of overall costs, the purchasing department is taking an increasing role in being responsible for achieving the required margins and profit.

Few, if any, companies in the UK have the same relationship with their suppliers as in Japan, where suppliers to the automotive manufacturers keep their stock circulating in lorries to ensure delivery to the minute. Neither is such a technique necessary for anyone but a volume manufacturer.

JIT supplier implementation will usually include the following:

- Building relationships
- Single sourcing
- 100% quality
- Regular, scheduled deliveries
- Delivery procedures.

Building relationships with supplier is vital in order to overcome the view amongst suppliers that JIT is only shifting the burden of carrying inventory onto them. This one-sided view, which is often shared by purchasing, must be broken down by education. Methods of vendor co-existence must be worked out with the ideal of sharing business objectives. The possibilities should be explored of suppliers themselves implementing JIT within their operations, and, in turn, with their suppliers. A supplier may have a similar cost distribution to one's own, as shown in Fig. 7.4. In many instances it is more extreme, with some high technology companies having material costs of well in excess of 80% of direct manufacturing costs.

Monitoring of supplier performance provides objective information to discuss during regular joint meetings. The supplier is expected to warn of any potential supply problems, and also expects to be warned well in advance of changes in demand. Many companies are implementing on-line communications with suppliers in order to eliminate paper work with all its associated delays and errors. Suppliers can also be involved at an early stage of design to help ensure the best design in terms of quality, ease of manufacture and overall cost.

Single sourcing is a natural outcome of good supplier relationships built upon two-sided trust. Dual sourcing may appear cheaper, but can be costly as parametric, functional or physical differences between vendors cause costly production or site failures. It is also symptomatic of the adverse 'us' and 'them' attitude.

100% quality eliminates the need for goods-incoming inspection. It will never be achieved when purchasing supplies on the basis of price alone, but is gradually achieved as suppliers are educated, and those who cannot, or do not want to, meet the criteria are eliminated. Technical appraisal and quality assurance of suppliers become two key factors in supplier selection.

Regular scheduled deliveries are important from the supplier's point of view, as this helps them to plan their own procurement and production. Delivery procedures are also important. Bulk components may be delivered straight to the line. Packaging methods and containers may be delivered to ease count verification, ensure material protection and simplify handling. Ideally, individual parts should not have to be touched until they are actually used for production.

The benefits to be obtained from improved supplier orientation are too important to miss. For example, over a four-year period one major UK manufacturer achieved the following:

- Number of suppliers: down from 1000 to 250
- Deliveries on schedule: up from 85% to 97%
- Goods-incoming defect rate: down by 12:1.

7.2.9 Summary
While this has served to review the manufacturing operations, lessons are equally applicable to other aspects of a company's operation. Paper-work systems, design methodologies and order-handling procedures can all benefit from JIT concepts. Indeed, the full benefit of shop-floor implementation will never be realised unless matched by similar measures in the office systems.

7.3 Computer systems

There is a great danger that, because facilities are available in the form of cheap computing power and data storage, these will be applied to try and resolve problems which should not exist at all. Here we refer to applying MRPII packages to plan, monitor and control production in place of restricting their application to planning. This stems from the blinkered view that electronics manufacturing is a complex process and therefore requires complex controls. The essential aim should be to simplify the process and simplify the controls. For example, take a medium-sized plant with an annual production of 200 000 PCBs and a production time of eight weeks. The task of controlling the flow of material, tracking, WIP, implementing design changes and so forth, with over 30 000 PCBs somewhere in production, is very complex. If JIT is implemented, the production time can be reduced to two days. The associated control problem is now reduced, with perhaps only 1500 PCBs now in progress. Gone is the need for complex computerised control and tracking systems or progress chasers. These are replaced by simple, visible and responsible management procedures, with the JIT techniques themselves imposing the control.

The production planning and control concept of the MRPII and JIT approaches are illustrated in Table 7.2 and Fig. 7.5, which illustrate the fundamentally different approaches to the tasks.[6] MRPII attempts to plan on a global basis, considering a large number of parameters and variables; it is therefore a very complex task. On the other hand, JIT is concerned with flow, and flow is determined by bottlenecks. Planning of JIT production needs to concentrate on how to ensure flow through the bottlenecks, as these determine the overall flow through all subsequent processes. Simulation can be used to identify the bottlenecks, and thus to provide the starting point for the organisation of planning and control. A number of systems and software packages are available which concentrate planning around the bottlenecks, based on synchronised manufacturing. This aims to maximise flow through the bottlenecks, preventing queues from building up.

JIT therefore offers a very much simpler approach to the task of control, primarily

Table 7.2 *MRPII and JIT: a comparison*

	Push	Pull
Organisational impact	Total company	Total company
Planning	Extensive	Limited
Computer systems	Complex, large	Simple, smaller
Simulation	Extensive	Limited
Control	Sophisticated	Simple
Parameters	Fixed	Variable
Scheduling	Push	Pull
Responsiveness	Sluggish—weeks	Fast—Hours
Summary	Extensive planning: Execution—Restriction and inflexible	Limited planning: Execution—Simple and flexible

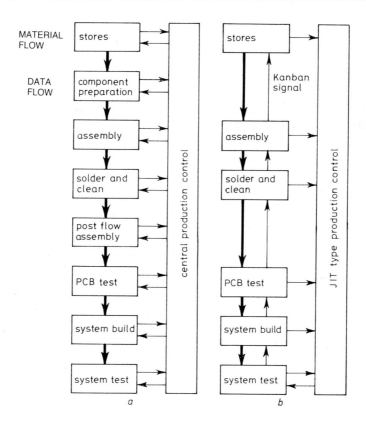

Fig. 7.5 *Traditional and JIT production: control principles*
    a Traditional approach
    b JIT approach
    ↓ Material flow
    → Data flow

arising from streamlined production and improved flow. For example, in Fig. 7.5 the tasks of component preparation and post-flow assembly have both been eliminated—the latter, by component re-selection, eliminating components which cannot be cleaned without damage and by product re-design. Component preparation, where required, is now carried out on-line. However, even this task is now much reduced by improved methods and elimination of solder-mask application. In the JIT environment, production is simply initiated by a pull signal into system test, with Kanban signals initiating action to the preceding stage. Information still flows out from the individual process areas to a central point providing the vital performance, test and inspection data. This compares with the traditional approach wherein material demand is referred back to a central point and action is initiated according to pre-planned and inflexible production schedules.

Inventory management is much simplified in a JIT environment, because the very short throughput times enable raw material and work in progress to be combined into a single category—raw and in-process material (RIP). Detailed tracking, no longer required, is eliminated. Inventory integrity is maintained by simply deducting the material list from RIP balances each time a product is completed, and having transactions to cater for the limited amount of re-work and scrap.

The JIT environment replaces complex and sluggish centralised systems with fast, responsive local control, basing decisions primarily on local and immediate knowledge and downstream demand. The interface with central control is now limited to final product demand and performance feedback data.

A major objective of JIT is to obtain a level rate of production with a flexible mix that reflects the market demand, with flow determined by true demand for output. An MRPII system which provides long-range capacity planning can now be used correctly, instead of being a tool to communicate and manage constant schedule changes.

JIT will not remove the need for an MRPII-type system; rather it requires a different emphasis. Referring to Fig. 6.2, all modules will still be present; however, as Fig. 7.5 illustrates, the shop-floor control module will be considerably simpler.

7.4 Summary

JIT is not a technique reserved just for the shop floor. It is applicable, and indeed should be applied, throughout a company. It can improve flow and quality—eliminating waste, irrespective of whether it is material on the shop floor or information in the office.

The concepts of JIT can be applied to virtually every production environment. However, the emphasis of implementation will vary according to such parameters as the nature of product and market. The implementation of JIT concepts in isolation will very rarely provide a universal solution. JIT will invariably be coupled with the implementation of technology-based techniques described elsewhere in this book. For example, total quality may well rely on the use of a comprehensive CAD system, engineering data bases and automatic assembly equipment. Simplification of the manufacturing process could be achieved through the use of advanced component technology, with one custom silicon chip replacing two PCBs, or it may involve the use of the sort of robotic flexible assembly system described in Chapter 12. The JIT concepts described are some of the key complementary steps needed to realise the full potential from investment in advanced manufacturing techniques.

7.5 Bibliography

1. ABE, K.: 'How the Japanese see the future in JIT'. Proceedings of 1st International Conference on JIT in Manufacturing
2. COX, J.: 'The goal' (North River Press)
3. DALE, S.: 'JIT and its impact on the supplier chain' in 'JIT, and execuive briefing' (IFS)
4. DALE, S.: 'Just-in-time beats just-in-case', *The Engineer*, 4 Dec. 1986
5. GOLDRATT, E. M.: 'Boosting shop floor productivity by breaking all the rules', *Business Week*, 25 Nov. 1984
6. HARTLAND-SWANN,: 'Just in time', *Industries Computing*, Aug. 1986
7. JACOPS, F. R.: 'OPT uncovered', *Industrial Eng.*, **16**, 10
8. LOWNDES, J. C.: 'Production management concept cuts delays and budget overruns', *Aviation Week & Space Technol.*, 13 May 1985
9. SIMPSON, A.: 'Effective JIT manufacture at HP' in 'JIT, an executive briefing' (IFS)
10. SUZAKI, K.: 'Comparative study of JIT/TQC activities in Japanese and Western companies'. First World Congress of Production Inventory Control, Vienna, Austria, 1985
11. SUZAKI, K.: 'Japanese manufacturing techniques: Their importance to US manufacturers', *J. Business Strategy*, Winter 1985
13. WARD, J.: 'JIT production—theory into practice' in 'JIT, an executive briefing' (IFS)
14. WILLIAMS, P.: 'Understanding JIT manufacturing', *Production Eng.*, July/Aug. 1985

Chapter 8

Support tools

This chapter addresses some of the key support activities and tools behind AMIE. Included is a discussion on communications, data bases, data standards, bar codes and fourth-generation languages.

8.1 Introduction

The preceding chapters have covered the basic areas of design, manufacture, test and production management. There is considerable benefit to be achieved through interfacing and integrating these together. This can help to ensure the accuracy and cohesion of data across a company. In addition, use can be made of powerful software tools in order to develop software to meet the particular requirements of each company.

There are many other peripheral aspects of AMIE to be considered, some which have been in existence for some time whilst others are still evolving. A basic understanding of all is required. This chapter will review some of these aspects, covering:

- Communications standards
- Data bases and their management
- Data standards
- Bar codes
- Fourth-generation languages.

One of the prime requirements for implementing computer-based systems, whether for design, manufacture or management, is to try to re-use information both within and across the systems. Some significant advantages are:

- The elimination of data transcription errors associated with manual data entry
- Data transfer can be virtually immediate and not subject to queuing delays.

The consequence of these advantages can be very far-reaching. Staff, who may often be highly skilled, can be released from mundane manual data entry, thus enabling them to be used effectively in carrying out their prime tasks. Provision of manufacturing and test programs directly from CAD should enable small batch automation, because on-line teaching-type programming will not be necessary and the debugging normally associated with manually entered information will be eliminated. This brings the volume break point at which automatic assembly or test becomes viable down to a very low level, and can provide a substantial saving, as the example in Chapter 2 illustrates.

8.2 Interconnection, interfacing and integration

Systems can be connected in a variety of ways, as Fig. 8.1 illustrates. Although integration is the ultimate goal, interfacing and interconnection can also both be worthwhile. The differences between these three lie in the manner in which the systems themselves are connected, and also in how and what information flows from one to another.

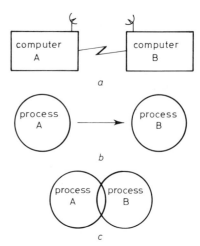

Fig. 8.1 *Interconnection (a), interfacing (b) and integration (c)*

8.2.1 Interconnection

Interconnection of equipment enables information to be transferred from point A to point B without the use of any transportable media such as discs or paper tape. It can be achieved through a variety of methods. Section 8.3 reviews some of these. The exact details of the interconnection method may be immaterial. For example, if a public network is used for inter-site communication, the end user will be unaware of whether data is transferred by wire, fibre optics, a microwave link or even satellite. However, although equipment may be interconnected, this in itself does not ensure that transferred data can be used meaningfully.

8.2.2 Interfacing

Interfacing allows data, usually specific and limited, to be transferred from one software package to another. The transfer process will usually be one way. The software packages being interfaced could be on the same computer, or they could be run on different computer systems.

Examples of interfaced systems are seen where a CAD system provides a parts list to an MRP system, or perhaps net and parts lists to an ATE system.

8.2.3 Integration

Integration, as Fig. 8.1c illustrates, is aimed at merging processes together. The operation of each process will still be separate, but data will be extensively shared and interfaces will often be transparent to the user. Integration of schematic capture and simulation packages with a PCB layout system will allow circuit information to be transferred to the layout system. Feedback from the layout system will back-annotate the circuit and allow re-simulation to take into account effects such as the copper track capacitance.

8.3 Communications and networking standards

To be successful in the use of AMIE, applications supporting a range of different tasks run on a number of different computers, which will need to be connected together. These computers will in the majority of cases be of different types. In practice, the task of connecting different makes of computer together, so that data can be easily, accurately and intelligibly moved between applications, is difficult.[2] The difficulties are increased by timing and addressing issues when more than two computers are involved and the computers are connected into a network configuration. Incompatible networking products from different computer manufacturers make the task of connection both technically difficult and expensive to achieve.

This problem can only be solved by the wide acceptance of international standards. In 1979 the International Standards Organisation (ISO) published a document describing how the problem should be solved. This document described a reference model for the organisation of the various tasks involved in communications and networking called Open Systems Interconnection (OSI). The model addressed more than data transfer and covered all aspects of applications to communication across a local or wide-area network. Progress towards the implementation of OSI is slow, but the necessary standards are being agreed and many leading vendors are committed to the support of OSI as an alternative to proprietary products.

The reference model is applicable to a wide range of situations other than manufacturing, but clearly manufacturing has some specific requirements. It was to address these requirements that General Motors began the development of the Manufacturing Automation Protocol (MAP) and Boeing the Technical and Office Protocol (TOP).[11] MAP and TOP standards are still evolving and their widespread use is still some years away. Once the standards stabilise, their importance will be very significant and any forward-looking strategy for the use of AMIE should consider the role of MAP and TOP.

Fig. 8.2 shows a simplified view of the 7-level OSI reference model. Fig. 8.3 shows the relevant standards for each of the 7 levels of the model as applied to MAP and TOP. It can be seen that they differ only at levels 1, 2 and 7.

At levels 1 and 2 the standards define broadband as the physical medium, and token bus as the media access method for MAP. This is to meet the requirements for bandwidth, speed and timeliness of data transfer, as well as to meet the environmental conditions of the factory. Alternatively, the TOP standards define baseband as the physical medium, and carrier-sense multiple access with collision

58 Support tools

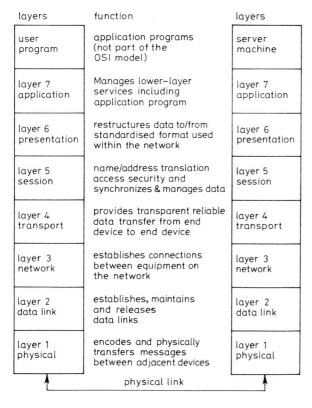

Fig. 8.2 ISO reference model

detection (CSMA/CD), as the media access method, in order to meet the simpler requirements of the office.

Layers 1–6 are concerned with the timely and accurate movement of data. Layer 7 is concerned with the format of the data and whether the dialogue takes the form of short messages or data-file transfers. Thus layer 7 is the area of standardisation of most interest to the user of AMIE.

It is intended that eventually data exchange-standards such as Electronic Data Interchange Format (EDIF) and Initial Graphical Exchange Specification (IGES), described below, will form part of the total definition of MAP and TOP. In the current release of the standards an area of great interest and relevance to AMIE is the Manufacturing Message Specification (MMS). MMS will form the basis of a set of data-exchange message standards for specific tasks involving communications to and from programmable devices in a CIM environment. A message-based dialogue between a cell controller and a numerically

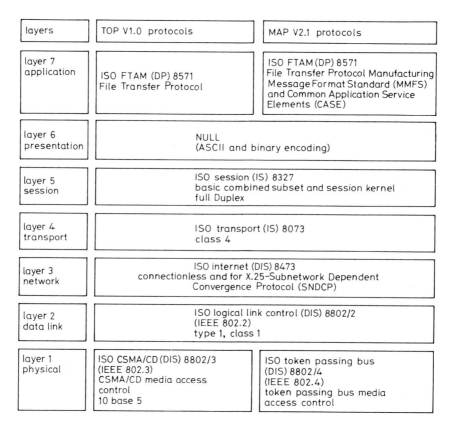

Fig. 8.3 ISO reference model with MAP/TOP common core of protocols

controlled component insertion machine is an example of such a task. In the design office environment the File Transfer Access and Management Protocol (FTAM) facilitates the transfer of data files between alien systems. For example, this could be the transfer of a parts list from the CAD to MRP system.

MAP and TOP are implementations of the OSI model applied to a local area. In many situations there will be an equally important requirement to communicate over a wide area. Multi-site companies or companies co-operating on the development of a project will have a need for data communications over a wide-area network, and this may also need to be built in to the AMIE strategy. This may extend to the need to communicate digitally to suppliers over a value-added network, particularly as closer relationships are built up with suppliers through the use of JIT techniques.

Besides the standards mentioned, there are other options for communications such as fibre optics or RS232. In the 7-level model these relate to lower levels only, and could therefore be used for that function. Illustrated in Fig. 8.4 is a possible communications network for a large site. This makes use of the various MAP and TOP options, together with RS232 and fibre optics as the level 1 communication media.

At the lowest level, the choice of communication media is governed by such factors as the volume of data transmission, required response, distance and security. Table 8.1 reflects some of the more important factors governing choice.

8.4 Data bases and data management

The adoption of advanced manufacturing systems results in information no longer being held as paper records, but as computer records. Storage of information by computers can be by one of two methods: a data-base management system (DBMS) or discrete computer files. The latter often suffers from the problems usually associated with paper-work systems, such as duplicate copies, misplaced information and different versions of a document. However the use of a DBMS can overcome all these difficulties and also provide further substantial advantages. The major benefit of a DBMS is that, as soon as an update is made to any data record, any subsequent access to the data base can readily make use of the updated information.

There are three main types of database—hierarchical, network and relational.[5] A hierarchical data base is, as the name implies, hierarchical in organisation. Information is expressed as a series of one-to-many relationships. In a network data base the relationships are expressed as many-to-many. Because relationships are built into the structure of these data bases, both types can offer fast access to information and efficient use of storage space. However, they are best suited to stable en-

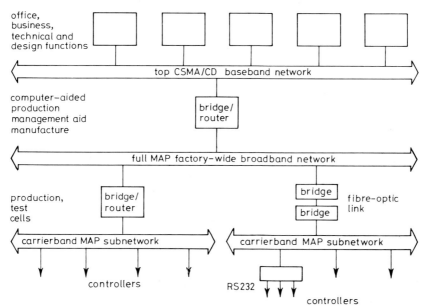

Fig. 8.4 *MAP/TOP network*

Table 8.1 *Comparison of communication methods*

Parameter	RS232	Base/carrier band	Broadband	Fibre optic
Speed (typical)	19 200	50 Mb/s	10 Mb/s	200 Mb/s
Cable Channel Capacity	1	1	hundreds	1
Distance	10 m–12 km	1·5 km	65 km	unlimited
Repeaters	modem as required	every 1·5 km	every 1·3 km	every 2·5 km
High data rate-graphics	no	no	yes	yes
Video conferencing	partial	no	yes	yes
Max no. of terminals	low	medium	unlimited	unlimited
Expansion (devices)	limited	some scope	few limits	few limits
Expansion (speed)	limited	limited	few limits	few limits
Data error rate	1×10^{-5}	1×10^{-7}	1×10^{-9}	very low
Noise vulnerability	yes	yes	no	no
Data security	low	moderate	high	very high
Ease of network mods	difficult	easy	very easy	difficult
Cable bandwidth	3 MHz	50 MHz	400 MHz	very high

vironments where the uses of the data base are determined in advance.

In a relational data base all data is held in a series of arrays or tables. This type of data base offers considerable scope for flexibility and data independence, because every table is independent of others and data in different tables can be combined or compared. No pre-existing relationships tie the tables together, and relational data bases are structured with no preconceptions about the manner in which the data will be accessed or used. The long-term flexibility of relational data bases makes them an ideal choice for many applications. In the field of AMIE they are often used for applications such as CAD systems and engineering data bases.

However, irrespective of the type of data base, the definition of structure, contents and fields is an important task. The needs of all possible users should be considered, together with the requirement for long-term flexibility. Inevitably, the data base will never be static, and its structure should be able to cater for change. Long-term growth must not be restricted by such data-base limitations as the number of data records, length of records and number of fields associated with an item.

The physical method of storing the information will depend on the type of computer system used. In a networked system, information may be stored in different physical locations, so that the user could be unaware of where the data is stored.

Irrespective of where and how information is held, the requirement for an effective data archive and back-up cannot be over-emphasised. Archiving is a means of off-loading information from the computer when that information is no longer immediately relevant. Back-up is concerned with security copies. Secure back-up procedures must be established to enable the computer to be restored to a recent known state following either accidental or malicious loss or the corruption of data. Archive and back-up routines will reflect the nature of the business and the extent of reliance upon the computer system. Many companies will keep a mirror image of all data, so that, when a transaction has been completed, automatic routines will immediately make a mirror-image back-up copy. Other companies may take a daily incremental back-up copy of all data which has changed that day, a weekly copy of changes and monthly a complete copy of the information.

8.5 Data standards

Fundamental to the success of AMIE is the flow of information from one process to another. While the development of communication standards, based upon the ISO/OSI model, has eased the mechanics of transferring the information from one process to another, it has done little to simplify the task of presenting that information in the correct format. It is here that data standards help.

Any company wishing to transfer data between systems is presented with a complex task of implementing and maintaining a multitude of individual links. As the range of systems grows the task becomes an unacceptable overhead, as Fig. 8.5 illustrates. It is here that standards help. Several overlapping standards are emerging as relevant in the field of advanced manufacturing. Work is going on to develop a standard known as Standard for Exchange of Product Data (STEP), taking the best from each of the individual standards. This is likely to be released in a usable form by the early 1990s.

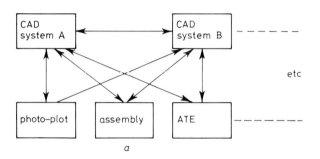

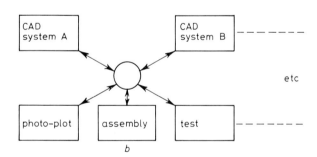

Fig. 8.5 *Data-exchange options*
 a Direct interfacing
 b Use of neutral interchange format

The major standards of current significance in electronics are as follows:

8.5.1 EDIF: Electronic Design Interchange Format
This is a *de facto* standard, and is not yet ratified by a standards organisation. It is, however, very widely supported. Initially aimed at integrated circuits, EDIF now aims to provide a single format for transfer of electronic circuit design-related information. It includes not only such aspects of design as net lists and physical layouts, but also behavioural descriptions.

Two particular attributes of EDIF make it particularly useful: the ability to support hierarchical design and the format of the data itself.[8] The syntax of the language requires keywords to be placed at the beginning of each list. Software accessing the data will skip any information it does not require or understand. Once the standard is stable, it will provide a powerful tool for long-term data archive and storage.

8.5.2 IGES: Initial Graphics Exchange Specification
IGES is a neutral intermediate format intended for transferring two- and three-dimensional graphical and related information between different CAD systems. Version 3 includes schematic and connectivity descriptions, together with physical layouts of PCBs and ICs. To date, the applications of IGES have primarily been in mechanical engineering rather than in electronics. Successful applications of IGES are limited where complex geometries, such as 3D curved surfaces are involved. This is due to the variation in data definitions and algorithms between different CAD systems.

8.6 Bar codes

Bar codes provide a ready means of accurately and rapidly entering data into computer systems. They offer a substantial improvement in speed of data entry and also in the integrity, format and standardisation of that information over other methods. For example, keyboard entry has an error rate of 1 in 3000; optical-character reading, 1 in 10 000; and bar codes 1 in 3×10^6 character reads, respectively.

Bar code symbols, of which there are many types, carry information in the relative widths of bars and spaces. For example code 3 of 9, as Fig. 8.6 illustrates, encodes each character with nine elements, where three out of the nine are wide and the remainder narrow. This self-checking code is the most widely accepted in the electronics industry. The standard printing density of code 3 of 9 is 9·4 characters per inch, but this can go as low as 1·4.

Bar codes can be printed in many different ways. For PCB applications pre-printed labels could be applied, or the code written directly

Fig. 8.6 *Code 3 of 9: character B*

by means of ink-jet printing. Alternatively, the code can be written on the edge of the PCB using a laser.

Reading of the codes can be achieved using either contact or non-contact systems. Contact systems, utilising a hand-held wand, are relatively cheap, but can damage the printed symbols and may be subject to dirt, thus making reliable reading a problem. Non-contact systems comprise either a fixed beam, with the label being moved underneath, or a moving beam. The latter, usually employing a laser scanner, are considerably more expensive than the hand-held wand, but do offer advantages because they scan the label several times, thus offering a very much higher degree of first-time reads. The choice between laser or wand is dependent on the application.

8.7 Fourth-generation languages

A large number of programming languages exist. These can be split into distinct generations:

> *First generation:* machine code. Used for entering both data and processing instructions into early computers.
> *Second generation:* assembler languages. A mnemonic extension of machine code, allowing the easier memorising and interpretation of coding instructions.
> *Third generation:* high level, procedural languages, such as Fortran, Cobol and Basic. These combine low-level instructions into statements or commands, thereby enabling a much simpler and more intelligible coding of computer instructions.

The ever-increasing demand for more sophisticated and comprehensive computer systems has constantly demanded more rapid and accurate software development. A number of productivity aids have been developed as adjuncts to the languages. These provide products such as packaged sub-routines and macro-libraries, report-generating tools, screen design and handling packages, prototyping tools, data-base management systems, file-management and -generation packages and project-management and -development methodologies. All such tools simply offer improvements over the basic languages. Fourth-generation languages (4GL) provide the next significant step forward.

Typically a 4GL should have the following attributes:

- Fast and efficient means of constructing a complete application through either interpreted or compiled code
- High-level non-procedural programming for application-logic code generation
- File definition and maintenance
- Screen- or form-based input and validation
- Report generation
- Query processing
- Communications interfaces.

4GLs are still in their infancy, and although many products exist in the form of spread sheets and data bases, these often fall short of the full definition. A true 4GL should enable the systems analyst to generate code directly from data-flow diagrams. Flow diagrams are generally entered directly onto the screen, with an expert system monitoring and directing the process. Use of a 4GL can have a dramatic effect on the computing activities of an organisation, the most noticeable of which are:

- Substantial improvements in the speed and cost of both software development and maintenance
- Easy creation of prototypes
- Reduced likelihood of software errors
- Easy modification of applications, thereby reducing the maintenance burden
- Reduced need for extensive technical expertise in development staff, with greater emphasis on business analysis skills
- A greater part can be played by users in the design and development process, which helps to ensure the end product really meets their needs
- Smaller development teams and shorter lead times reduce the difficulty and complexity of project management control.

4GLs ease the task of translating system requirements into software. Use of such tools can make substantial inroads into traditional software development. However, they are not a substitute for high-calibre personnel capable of defining system requirements. The benefit of 4GLs is often obtained at the expense of the efficiency of the final machine code generated. For the majority of applications this will be acceptable; however, a lower-level language may be required where requirements such as speed are critical.

8.8 Bibliography

1. ALLASI, D. C.: 'Bar code symbology' (Intermec Corporation, 1984)
2. BLYTH, G.: 'Understanding factory LANs', *Electronics Manufacture and Test*, 1987
3. DRINNAN, D. W.: 'Bar code control of PCBs'. Proceedings of EPEE, 1985
4. DWYER, J.: 'Unravelling communications'. FinTech 4, Automated Factory
5. EMERSON, S. L.: 'Database for the IBM PC'. ISBN 0 201 10483 0
6. MICHAELS, E. S.: 'Qwerty versus alphabetic keyboards as a function typing skill', *Human Factory*, 1977, **13** (5)
7. HARARI, S.: 'Networking offers process paybacks', *Electron. Manufacture and Test*, Sept. 1985
8. KAHN, H. J.: 'An EDIF parser'. Proceedings of CADCAM, 1986
9. LAWRENCE, A.: 'CIM architecture', *Industrial Computing*, Sept. 1988
10. MARRIOTT, M.: 'Bar code systems and their application in manufacturing and distribution environments'. Proceedings of EPEE, 1985
11. MORGAN, E.: 'Through MAP to CIM' (Department of Trade and Industry, 1986)
12. RANKY, P. G.: 'Computer integrated manufacturing' (Prentice Hall, ISBN 0-13-165655-4)
13. 'CIMAP event guide' (Findlay Publications, 1986)
14. 'The OCR wand' (Recognition Products, Inc.)

Chapter 9

Implementation

This chapter discusses the method of implementing AMIE. It outlines the total process from developing an effective strategy, through planning, modelling, simulation and installation stages, to the final evaluation of the operation and its effectiveness against the original plan.

9.1 Introduction

Investment must always be carried out on a secure foundation. Intuition can point towards the general direction in which a limited amount of investment may give benefit. However, investment resulting from strategic business planning will undoubtedly provide a greater likelihood of success.

To illustrate this point, consider investment in CAD. The design department purchases a CAD system in isolation. They do not consider in any detail the benefit which CAD may have on other functions in the company. The manner in which the system is set up in terms of, for example, procedures and data-base customisation is oriented towards how the system operates rather than what the company requires. A number of problems can result. For example, a major benefit of CAD should lie in the control which can be exercised over component selection. However, this facility lies idle, design engineers are free to choose components at will, rather than considering some of the company-wide aspects of choice such as quality and manufacturing procedures and costs. Automatic assembly equipment cannot be used unless layout requirements are adhered to. Following a fragmented and 'sloppy' evaluation it is found, for example, that the installed CAD system cannot be used to ensure automatically that requirements are met. A manual checking process is required, followed by yet another design iteration.

This example serves to illustrate the problems with investment carried out from the 'bottom-up'. Top-down planning followed by bottom-up implementation offers considerable advantage over that approach. It would, in the case cited, have identified the scope and detailed requirement for building in manufacturing requirements, thus effecting the selection criteria and eventual choice of CAD system. Procedures and data-base construction would have been centered on what the company required rather than what the system could actually do. CAD could have become a key element in reducing the overall cost and development time-scale. Instead it was relegated to an automated drawing board, now incapable of providing all the potential benefits.

9.2 Initiating action

Investment can be initiated from a number of sources. For example, comparison of company performance against others will indicate the potential for action; or action can be initiated as a result of identifying new market opportunities. Irrespective of the starting point, investment should only arise out of sound strategic plans. An outline of the manner in which this should be carried out is illustrated in Fig. 9.1. Planning starts in the market place, identifying the potential and the goals which the company wishes to achieve. From this a strategy can be developed. This is not a detailed plan, but rather more the general framework and direction in which the company will move. It should consider resources in terms of people and finance and the available technology.

The resulting strategy can now be used as a blueprint for the detailed planning stage. This comprises four main elements; modelling, simulation, justification and evaluation. In the ideal world these actions would be sequential; however, in practice, they are all closely

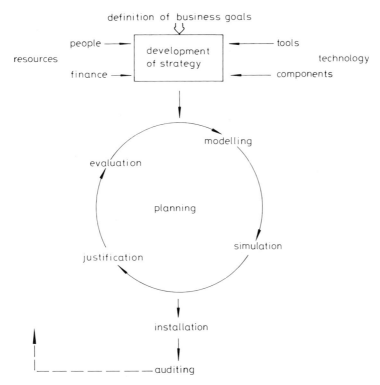

Fig. 9.1 *Strategic planning flow chart*

related, and thus the detailed planning stage is invariably iterative.

Once the decision has been made regarding investment, implementation can be initiated. Implementation covers far more than just 'bolting down the equipment'. It will include, for example, the people aspects of training and education, company procedures and building up the required data bases. Finally the performance of the system must be monitored. Planning stages should have produced realistic performance projections against which the actual performance should be checked.

The exact approach taken will vary from company to company. The following action-plan checklist is intended to highlight the major steps which a company should go through. It will be noticed that, of the 12 activities which are listed, half are directly associated with people—establishing teams, consultation, communication and commitment. Technology cannot succeed without people.

Action plan check list
- Recognise the seriousness and urgency of the need and identify the potential for action.
- Appoint a senior executive to direct a task force. He should be enthusiastic, committed and have authority.
- Establish a multidisciplinary core team with time, resources and authority to undertake preliminary planning and investigation. Define their objectives.
- Give Board level backing to the task force in all functional areas.
- Establish procedures for employee and trade-union consultation and communication.
- Develop, approve and back a strategic manufacturing plan, setting targets for next 3–5 years.
- Map out the way in which the company operates, showing how functions and departments interrelate, and identifying communication channels and requirements. Simplify the organisation by the integration of functions.
- Evaluate equipment and justify the solution.
- Draw up a detailed plan covering equipment installation, education, training and skills updating.
- Implement the plan.
- Audit the performance of the system in relation to planned goals.
- Give continuous, visible top-level management commitment.

Success of implementation is dependent upon many factors, not least the enthusiasm and commitment of the leader. Continuous and visible high-level management backing throughout is an essential if people at a lower level are also to feel committed.

The time and resources required to implement must not be under-estimated, and staff must be given the required resources. It is often necessary to ensure that key people are relieved of many of their normal duties during the initial planning and investigatory phases. The use of external consultants can often prove to be beneficial, either in supplementing scarce internal resources or in providing additional skills.

As important as the top-down leadership and commitment is the bottom-up co-operation. Employees can rapidly become alienated if they feel in any way threatened or exploited. Consultation with trades unions, and regular communication with employees over progress and plans, are both vital parts of the process.

9.3 Business goal definition

The profitability of a company is intimately related to performance in the market place. Therefore planning could start by identifying the goals which the company is trying to achieve in this area, rather than by asking questions such as 'how many engineering work stations can the company afford over the next 12 months?'

Market research may be required to ascertain market sizes, trends, state of the competition and what features the customer actually requires. Business goals can cover a wide range of performance-related factors, and typically would include a selection from the following:

- Company growth targets
- Change in profit margins
- Share of existing market
- Development of new markets
- Improved marketing performance
- Decrease in product cost
- Changes in product variety
- Changes to product specification
- Improved response to change
- Decrease in design cycle time
- Decrease in production time cycle
- Increase in production-volume capability
- Improvement to product quality and reliability
- Improved service to customers.

Laying down the foundation for developing the business is vitally important, and the various options must be developed and explored. The obvious may not always be the ideal. For example, a reduction in profit margin could, for some products, result in a considerable increase in sales, with the net result of increased turnover and overall profit.

Such decisions cannot be taken lightly, but must be the result of a thorough and sound basic approach. However, the definition of business goals need not be a lengthy process and the resulting document can be short and sharp. For example, one major UK electronics company, manufacturing large complex electronic systems, outlined its business goals in a very simple fashion:

- 10% cost reduction per annum
- 15% productivity improvement per annum
- Double the asset performance by 1990
- Improved quality and reliability equal to the best possible
- Provide 100% service to the customer—whether internal or external
- Provide a flexible and responsive service.

By aiming at these goals this company turned itself around in a period of two years. It has, for example, a delivery lead time of two months, as opposed to the competitors' average of six to eight.

At this stage the plan is purely limited to goals, and no mention is made as to how they might be achieved. For example, the 15% productivity improvement in manufacturing could be achieved in a number of ways—improved manufacturing quality, additional auto-insertion or improved design for manufacture are all possible solutions. Definitions of how to achieve results at this stage are liable to limit and constrain the next stages and result in sub-optimal performance.

9.4 Strategy development

Following completion of the business-goal definition stage, the development of a strategy is not a detailed plan, but is the general direction in which the company will move in order to reach its goals. It should provide clear guidelines for the company operations and developments over a considerable period of time, usually in the range of three to five years.

Development of the strategy is not a straight

through process, but is iterative in nature, objectively considering the options and selecting the optimum. Strategy is obviously concerned with the whole company. For example, we have already seen that changes in the design department could provide a major contribution to an increase in productivity.

Thus strategic planning will examine, attempt to understand and challenge functional relationships in the company. It will cover such functions as design, manufacturing, accounting, purchasing, marketing, aiming to integrate and steamline the operation of the company wherever possible. Questions will invariably be asked which challenge basic company structures and operating methods. They should identify the root of the problem or inefficiency within the company, rather than suggesting a means of resolving the symptom. This strategic planning stage should not get down to the very fine detail of planning. However, during this high-level planning stage, and especially in understanding the operation of the company, the use of modelling techniques should be considered. Some of the higher-level techniques described below can be useful.

This phase cannot be carried out in an isolated fashion, but must consider the availability of resources both in people and finance. Furthermore, it must be related to the practicality of the available technologies in tools and components.

The strategic plan will create the overall framework in which the company will develop. It should address the basic business issues and should provide clear directions on such matters as:

- Product volume, variety and specification trends
- Relevance of company-organisation structures
- Impact on people, educational requirements, re-training or the need for new skills
- Outline requirements for capital equipment
- Commitment to advanced component technology
- Make or buy issues
- Broad implementation time-scales.

To illustrate the outcome of the strategic planning phase, consider the following company. It designs and manufactures a wide variety of electronic warfare systems in small volumes. It has four major divisions all designing products with virtually identical electronic specifications. However, similar products are packaged differently to suit the varied requirements of land-based, mobile, airborne and marine applications. Each system contains 30–40 different printed circuit boards. Their definition of business goals included:

- Reducing new-product-introduction time scales
- Decreasing cost
- Increasing product reliability
- Improving functionality and performance of the product.

The similarity of the electronic circuitry in the products was identified and methods of exploiting this were examined. It was realised that a central core product could be defined. The resulting larger volumes justified the use of custom silicon rather than off-the-shelf integrated circuits. This in turn provided substantial benefit:

- Use of custom silicon increased reliability and could provide the means of improving the functionality and performance of the product.
- Costs were reduced as a result of being able to spread design costs over a greater volume of product.
- New products could be introduced by using the standard core of custom silicon-based PCBs. Custom silicon also demanded CAD, giving a right-first-time capability, eliminating many costly and time-consuming design iterations and production changes.

Had the company not approached the task strategically a different view would have emerged. For example, because of the lower volumes, the use of custom silicon would have been precluded if the goals had been applied to a single product only.

A sound, well investigated and well thought-out strategy backed by accurate data is a vital prelude to investment. It can prevent a fragmented and piecemeal approach, which will inevitably result in sub-optimum performance, potential compatibility problems and either a reluctance or inability to invest further owing to inadequate returns. Furthermore, it will avoid an addiction to technology, where technology is implemented for its own sake rather than to resolve basic business problems.

9.5 Modelling and simulation

The consequences of incorrect or ill-founded decisions made during the strategy development and planning stages can be devastating in nature. There is therefore a pressing need carefully to investigate and assess the operation of a new facility before investment takes place. A number of techniques have been developed which allow a model of the system to be constructed and evaluated.[10] Simulation allows the operation of the model to be examined. Parameters can be varied in order to assess, for example, the effect of batch size, and hence define the optimum size, or to consider the effect of order-input variations, and hence the most effective manner of handling them, or to examine the number of machines required to produce a given output.

Models can, of course, be constructed in a number of different ways. They include written descriptions, hand-drawn diagrams, physical representations and computerised techniques.

User-friendly graphically oriented computer-based systems present planners and analysts with some very substantial advantages over more traditional methods:

- They allow a complex operation to be decomposed into small, easily understood and well defined functional entities.
- They allow the model to be readily validated.
- They allow a ready understanding of the problems, system and solutions.
- They help the analyst to arrange a diversity of facts in an orderly manner, and also help to guide the analyst towards a complete analysis of the system.
- Once defined, computer-based systems allow system parameters and input data to be changed interactively in order to evaluate the effect of these on system operation. Many iterations may be tried in order to define a satisfactory solution.
- A hierarchically decomposed model allows understanding, analysis and investigation at any level. This allows the same model to be used to both develop and verify outline strategies and analyse detailed operation.
- Modelling and simulation will help clearly to identify potential problem areas and bottlenecks at the planning stage. This allows alternatives to be formulated and evaluated, in order to either eliminate or reduce the problems.
- Modelling and simulation help to identify the areas for investment which will produce the greatest benefit.

Many different tools are available. They split into three major types:

- Function modelling
- Information modelling
- Dynamics modelling or simulation.

It is unlikely that an ideal solution will be reached immediately. Invariably modelling and simulation tend to be iterative, continuing until there is no major benefit in further evaluation. Fig. 9.2 outlines a typical simulation project, indicating the major steps and illustrating the iterative nature of the task.

The main objective of using these tech-

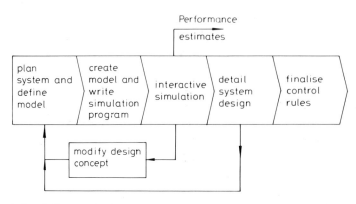

Fig. 9.2 *Outline of a typical simulation project*

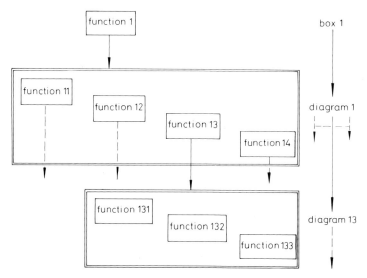

Fig. 9.3 *Functional modelling: decomposition*

niques is to help define a solution that will provide maximum benefit from an investment. Economic justification, discussed in Chapter 2, will require detailed information regarding throughputs, machine times, labour and so forth if the best system is to be found. Output from the three techniques mentioned can be linked to economic analysis.

9.5.1 Function modelling

Function modelling allows a complex operation to be broken down from a top level. Function modelling can be concerned with almost anything, such as company operations, system requirements or a technical system. The operation is decomposed, as Fig. 9.3 illustrates, until the lowest level shows all the detailed functions and interfaces. A functional modelling tool, such as the one illustrated, has a well defined structure to each block, as Fig. 9.4 shows. Each block, or function, processes the inputs according to the control mechanism and the available resources, in order to provide output. Both input and output can consist of material and/or information. Resources tend to be people or equipment, whilst control is often related to operating instructions or work schedules.

Shown in Fig. 9.5 is a functional model of a typical PCB test area. This shows the interrelationship of the tasks of organising the work area, setting up the test equipment, actually performing the tests themselves and carrying out re-work where necessary. The flows of information and material are also defined. Some of these interfaces may be restricted to

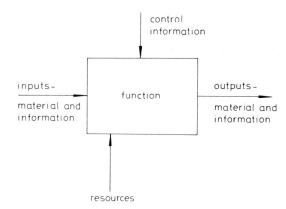

Fig. 9.4 *Typical functional-model-diagram representation format*

within the diagram itself, while many will flow to other diagrams within the overall model.

Functional modelling tools should provide syntax and model-analysis facilities to ensure a consistency between diagrams across the model and to check that the appropriate inputs and outputs are linked.

9.5.2 Information modelling

This technique is aimed at producing a model which represents the logical structure of the information to be stored. This will provide the system designer with the conceptual design of a data base which should be able to be mapped onto the majority of data-base systems.

A variety of tools exist to help in information modelling. They may exist in isolation, or some may be compatible with function-modelling tools. This allows the two models to be cross-referenced. It offers considerable advantage because it helps to ensure that the

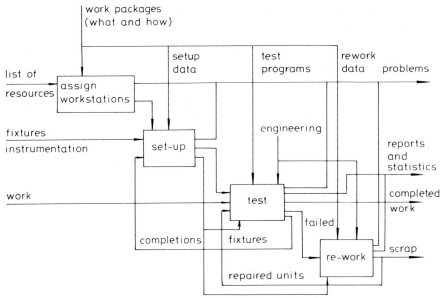

Fig. 9.5 *Typical functional model for a PCB test area*

data-base definition accurately matches the functional model.

9.5.3 Dynamic modelling

Dynamic modelling, often referred to as simulation, is concerned with time-dependent aspects of manufacturing operations. While the two former tools have been of primary interest to the computer-system builders, this technique has more relevance in analysing shop-floor operation. Simulation allows the designer to evaluate the effect of variables in order to produce an optimum design before commitment to the cost of detailed design or physical implementation. Fig. 9.6 illustrates a typical simulation process.

Consider, for example the design of a robotic-assembly work station. Many questions have to be answered—should there be two robots, one to feed and prepare components and the other to insert, or would a better solution be to have only one robot? If the robot fails to insert a component first time, how many times should it attempt to 'find the hole' by a jiggling technique before straightening the leads or rejecting the component? What is the optimum number of feeders to use? On what basis should parts replenishment be carried out?

Answering such basic questions as these is impossible to perform manually. Interactive simulation provides a powerful and ready means of answering such questions and also for projecting ahead and answering 'what if?' type questions concerned with work loading, order variances etc.

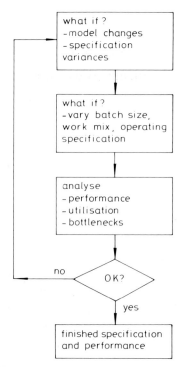

Fig. 9.6 *Simulation-model analysis*

9.5.4 Selection of planning tools

In order to achieve the optimum benefit from the planning tools, they must be chosen with care.[9] Their selection will require just as much care as that of capital equipment items. The following list covers aspects which are of importance when evaluating modelling and simulation tools:

- How are models built with this package, how skilled does the user have to be?
- Is there a high level of interaction available during model building and simulation?

- How large a system can be portrayed graphically on one screen?
- Are there limitations to the size and complexity of the model or to the run time of simulation?
- Can hierarchical models be constructed?
- Are there facilities for automatic syntax checks and cross-referencing between sheets for large models?
- What limitations are there to building complex logic functions?
- Are the graphics animated in simulated time or do they 'play back' after simulation runs?
- What quality graphics?
- What is the documentation like? Clear, easy to follow, sufficient detail?
- What vendor support is provided. What future developments are planned?
- What training is available?
- How portable is the software; what type of hardware is required?

Companies requiring to use simulation on an intermittent or one-off basis may not need to be fully conversant with the details of the technique. They may find it more effective to use the services of an external consultant who is fully familiar with a particular field of application.

9.6 Evaluation

Definition of strategy, planning, modelling and simulation will have brought the implementation process to the point of evaluation, in which systems, software, equipment or machines are evaluated in order to find the one which best matches the requirements. By implication, evaluation can only start once there is a clearly defined set of requirements, preferably ranked in importance. Evaluation should come after the basic justification has taken place and outline approval has been given. It should not be an unduly lengthy process, because delay at this stage means that the clearly identified benefits are not being realised.

Evaluation must be carried out objectively in order to provide recommendations which will lead to the correct investments being made. There is a very great danger of not defining the requirements until after a comprehensive vendor presentation. The 'requirements' are liable to reflect what the vendor offers rather than what the user needs. This may adversely affect choice. For example, one CAD requirement list, generated after a visit to a vendor, included the need for two screens per work station—one for graphics and one for command entry. This immediately eliminated any multi-window systems. Visits to users and vendors may form an important part of the early strategy development phase during which ideas are formulated; however, the definition of requirements must remain objective.

Requirements must be agreed which will meet the needs of the company as a whole, and not just the department directly concerned with the purchase. Furthermore, the interdependence of items needs to be carefully assessed. Consider, for example, the requirement check list generated by the design department for a PCB CAD system, and illustrated in Fig. 9.7. Take the general requirement concerned with the amount of system memory. Which is 'better'—a system with 512 kb or 5 Mb? In isolation the latter is better, but when considered in relation to other parameters this may change. Relate it to, say, the auto-router. The former system works on a 25 'thou' grid, whilst the latter is gridless. Immediately the question of system memory is invalidated, because of the fundamentally different approaches to routing, each with significantly different computing and memory requirements. Take as another example the question of manufacturing output. There are few CAD systems without a 'manufacturing output'; however, there may be very few CAD systems which can, for example, provide output for a specific surface-mount-placing machine. Fundamental questions such as 'can the manufacturing requirements be incorporated in the data base?' have also been excluded from the checklist. Requirements must be well thought out, far-reaching and clearly defined. They must also be defined before purchase, and not several months after installation has taken place.

A critical and objective approach to evaluation is not limited to CAD alone, but is equally important for all other items. Take, for example, the choice of auto-insertion equipment. Selection must go far beyond just considering the insertion rate and capital cost. Evaluation which covers the wider aspects of integration and flexibility is required. This will help to ensure that investment will achieve not

72   Implementation

General requirements

Display resolution/colours
Database access
Manufacturing and test outputs
Training, support
Pan and zoom
High performance processing
Graphics display

Operating system
Peripheral devices
Mass storage
Input device
Networking
CPU speed
Memory size

Component definitions

Component footprints and pads
Rotations
Alternate symbols
Ease of updating

Vendor's component library
Number and type of components
Accuracy and completeness of data
Layer conventions

Component placement

Automatic placement
Automatic optimisation
Gate/pin swapping
Interactive

Dynamic rat's nest/rubber-banding
Node highlighting
Placement by type, area, signal
SMD

Routing

Automating
Basic algorithm
Multilayer
Gridless
Rip-up and re-route
Finishing/via minimisation
Specific support:
– Analogue
– Surface mount
– Memory arrays
– ECL

Interactive editing
True geometry displayed
Ease of via creation and layer change
Curved angle traces
On-line/batch design rule checking
General drafting capabilities
Polygon fill for ground planes
Cross hatch, solid
Split copper planes
Limits:
– Number of components/connections
– Board size/dimensions

Fig. 9.7 *CAD-specification check list*

only today's requirements but also will provide an assembly facility that will give the flexibility and scope to deal with tomorrow's. Apart from the normal issues of speed, board size, cost etc., the points below are typical of the wider aspects which need consideration. A list such as this will need to be created by input from many different functions within a company:

Programming: Is the equipment capable of off-line programming? Are there any host facilities available to help the task? Can CAD-produced data be used without modification? Can the vendor accept standard data formats such as EDIF?

Data storage and communications: How much memory capacity is there, not only for storage of the current program, but also for alternatives? Can the equipment link directly to other systems, or must all communication be by a floppy disc?

Control system: What hardware and software are used; does it utilise industry standards; is it expandable; has the user ready access to it either to modify or add facilities? How does the manufacturer update the facilities provided—indeed can the facilities be updated?

Operation: Can full remote control be achieved including remote monitoring of status information? Are industry standards adhered to; e.g. the Manufacturing Automation Protocol—Manufacturing Message Standard (MAP–MMS)? What degree of sophistication is built in to allow automatic error recovery?

Vision: Is it provided; does it slow the system down; how easy is it to program; does it readily cater for component alternatives?

Accuracy: Has the system sufficient repeatability accuracy to cater for design requirements? Will design tolerances such as hole sizes have to be relaxed to an unacceptable level? Is the absolute accuracy of the system high enough to remove the need for manual debugging of automatically generated programs?

Component feeding: Can the system be equipped with enough feeders to cater for a large single product with many different types of component? Can it hold on-line enough parts to cater for a variety of products? What are feeder set-up times? For volume applications can alternative feeders be automatically selected whilst idle ones are re-loaded on-line? What is the cost of providing sufficient removable feeders to permit off-line set-up?

Board handling and fixturing: Can this be automatic? Can conveyors and magazines readily interface with equipment from different suppliers?

Component verification: Can component verification take place prior to insertion? What is the quality or depth of the verification test? What effect does it have on insertion rate?

Re-firing: Can the equipment automatically detect an assembly failure and problem? Can it re-fire and recover from the problem? Is there any user control on the number of attempts on one component? Is there full reporting of any problems in this area?

Footprint: How much top-side board area is required to cater for the insertion head? How much bottom-side board area is required to cater for the cut and clinch mechanism? Does this unduly restrict product specification and design?

Evaluation tends to be an iterative process, and a typical simplified route through this is illustrated in Fig. 9.8. Desk research will attempt to obtain as much information as possible. It will encompass vendor data sheets, case studies and articles. This will be sifted and sorted against the requirements in order to produce a short list of, perhaps, five or six vendors from an original list of, say, 15 or 20. It may be necessary for team members to have one or two presentations during this period in order to help them relate theory to practice.

Detailed presentations should be made by

Fig. 9.8 *Alternative routes through evaluation*

the vendors contained in the first short list. These presentations should include not only the technical details but extend to a view of the company, covering growth, size, installed base, support, training etc. Costs will need to be discussed in some detail at this stage. Systems should not be summarily discounted at this stage on cost alone without a clear definition of what is being provided. An apparently cheap system can end up costing far more than one which appears at first sight to be more expensive.

Evaluation should now have reached the stage where a second short list can be drawn up. This should consist of no more than, say, three vendors, and may already have been reduced to two or even one. User visits are an invaluable way of getting a broader view of the vendor company and assessing the operation of the system. Vendors are often keen to be present at such meetings, and a balanced view should be taken as to the desirability of this. Key points to be raised during user visits will cover support, reliability, ease and method of use, limitations and the overall benefit which the system will provide.

Bench-marking, i.e. assessing the relative operation of systems in a structured manner, often forms one of the last stages of the evaluation process. There are diverse views concerning bench-marking. Some entirely discount the need for it whilst others propose bench-marking eight or ten systems. A realistic approach can be to perform a single bench mark as a confirmation of the previous work. Alternatively, a comparative bench mark of, say, two of three systems can be carried out.[4] Bench-marking of many more systems is often an indication that the previous stages have not been properly carried out, or that parhaps the evaluation team is losing direction and objectivity.

As with the rest of the evaluation process, bench-marking will be of little value unless the objectives of the bench mark and the method of operation are both clearly defined from the outset. A CAD bench mark does not have to be centred on a complex design. For example, one company has devised a very simple circuit, utilising a handful of ICs, which will demonstrate the capability and limitations of an auto-router in a very short time. The result can readily be analysed and can provide far more useful information for the evaluation process than to be told that a PCB with 120 ICs on it took 47½ hours to route. In this instance, what trade-off is there between time, via holes, PCB layers etc. There is also a great danger that the bench mark will fail to demonstrate the capability of the system, and it can easily become a test of operator skill, stamina and resourcefulness.

A bench mark may be very time consuming. For example, circuit data capture and PCB layout for a large design may take two or three weeks, and it is usual practice that the bench mark is monitored for all of this period. This will provide valuable insight into how the system really operates, its power and its limitations, as well as ensuring validity and an understanding of results.

9.7 Preparation and installation

Implementation of advanced technology will invariably have far-reaching consequences. For the implementation to be a success, much preparation work is necessary prior to installation. The various stages of business definition, strategy development, justification and evaluation will all have considered the people aspects and the impact of the technology on the company. All this needs to be drawn together in a single coherent plan which will take the whole company forward. Such a plan is illustrated in Fig. 9.9.

As the goals are defined and strategy worked out, the changing roles of people will start to be identified, and at this stage education can start. This education will be aimed at changing attitudes and traditional ways of thinking. The exact changes that will be required will obviously vary with each individual company. However, several key attitudes which require changing are common to many companies:

- Functional integration
- Quality
- Waste.

Functional integration is aimed at identifying the interaction between departments, e.g. design and manufacture, and increasing the level of co-operation until the demarcation that exists is broken down. Attitudes such as 'the company cannot survive without this department' have to be faced up to and revised to 'this department cannot survive without the rest of the company'. The 'not invented here' syndrome must be counteracted. A good example of functional integration is cited in the second case study in Chapter 10. In that example design and manufacturing engineers change roles every six months, thus providing the cross-fertilisation of ideas and methods necessary to foster functional integration.

The realisation of 'right-first-time' operation is vital in any implementation. It is not a requirement peculiar to just-in-time manufacture, neither is it restricted to production alone. It must be an attitude and practice that is positively engendered throughout the company. People at all levels should be encouraged to have a pride in the quality of their work.

The identification and elimination of waste will also feature in many plans. The concept of waste elimination can be emotive because it pre-supposes that waste is currently tolerated. Many long-standing procedures will be identified as wasteful, and these have to be identified and either changed or eliminated. Some of these practices can be deeply engrained, and may be reflected in the total management style of a company.

As implementation progresses, the emphasis of education will shift from changing basic attitudes to specific training in new skills. This will start with the selection of people to be trained in the new technologies. Care must be taken to ensure that candidates have sufficient aptitude for the task, and are both willing and able to adapt. Existing computer literacy will often help in cutting down on the initial learning time, but is not an essential attribute. Training will usually be carried out by the system vendor. However, vendor training is not always sufficient because it will tend to suffer from the following faults:

- Too much emphasis is given to basic machine functions.

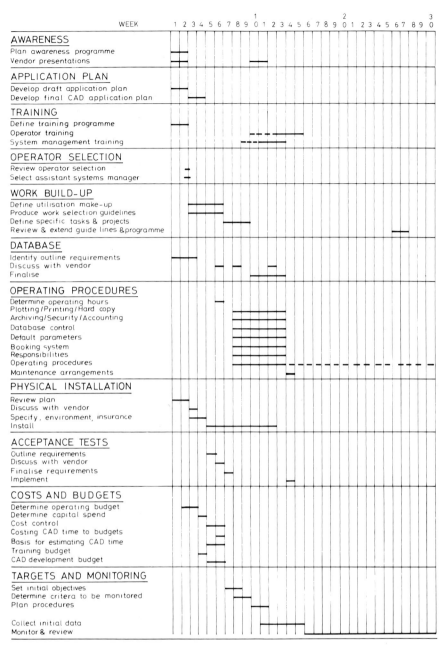

Fig. 9.9 *Typical CAD implementation plan*

- Too little emphasis is given to a more creative approach to design.
- Examples are too generalised and are not always relevant to the specific application.
- The training itself may be unstructured and unambitious in terms of performance improvements.
- Too little training is built into the software itself.
- Staff who will make demands on the facility without themselves operating the work station are ignored.

For example, a CAD training course will show how to create shapes for PCB layout, but it will rarely deal with the specific requirements for a particular company. A computer operating system training course will show how to carry out back-up and recovery procedures, but it is unlikely to discuss which procedure is most applicable to an individual company. It must be realised that training is only the start of getting going.

Installation of equipment should be scheduled to take place with training. Consideration should be given to the supply of services, access, floor loadings, environment and the like. In the case of CAD, thought should be given to the logistics of operation and methods to limit noise and external disturbance, whether it be from printers, plotters or an adjacent corridor.

An important part of planning will be to consider how best to utilise the equipment in the early stages, and how to build up work. With CAD there may be a temptation to use the system to design a complex product. A better approach is often to design a series of simple products to evaluate the complete set of operating procedures, rather than to get 'bogged down' in a complex component placing and routing problem. For automatic-assembly equipment a Pareto analysis of production volumes and component counts for various products will indicate the optimum products initially to target. Such a selection may be affected by whether the program can be generated automatically from the CAD.

The introduction of new systems will often force changes in procedures, operating methods and standards.[7] Some of these may be apparently quite minor, whilst others will be concerned with fundamental operations. For example, at the lower end, company standards such as drawing symbols will probably be examined. Although the CAD system should be able to cope with virtually anything, there may be advantages in particular approaches to the size and positioning of text, connection points etc. At a more fundamental level the responsibility for tasks will have to be reviewed. If CAD is integrated with ATE, with simulation output used as the basis of automatic test programs, the responsibility for the test programs may now shift from test to design. Once again, this will require matching with education programmes if implementation is to succeed.

9.8 Auditing

Implementation does not end with installation, but is a continuous and evolving process as the ultimate aim is constantly pushed further back and consequently rarely achieved. It is always possible to improve quality, increase efficiency or further reduce waste.

It is important that as operation is measured against the targets, any variances are identified and the underlying reasons analysed. Many parameters, such as WIP levels, throughput times, yields etc., can be readily measured. There are other factors which will be more subjective, and many of these relate to the human aspects—for example, how do people feel about the new practices? Are they understretched? Have they had sufficient training?

Having completed an assessment of performance, it will be necessary to take any corrective action that may be needed. For example, within the CAD area the task of PCB designs may be taking too long. Further investigation reveals that there is undue pressure to 'get on with the design'. Consequently the vital component placement stage is rushed through, with the net result that putting in the last one or two percent of tracks after autorouting has been completed is virtually impossible. Some specific on-the-job training in this instance helps to resolve the problem, as does education of management in terms of the approach to design using CAD and the need to spend time at the beginning in order to save it later on.

Auditing is not to be seen as a 'witch hunt'; rather it is a very constructive process. It is required in even the best of companies. It can enhance performance and thus help ensure future profitability, viability and investment.

9.9 Summary

The total implementation process—starting with defining goals and ending with auditing performance—may seem lengthy. Such an approach is not new but is good business practice. What is new are the powerful computer-aided tools which are available to model and simulate performance and operation.

The prime objective of all such techniques is to minimise the cost and risks, whilst maximising benefit and performance involved in complex situations. This is achieved by allowing the mistakes to be made in a 'safe' environment. No one argues about the sense in using simulators for pilot training or nuclear-power-plant operations; their value should not be questioned when determining how to maximise the return from a substantial investment.

9.10 Bibliography

1. BOER, C. R.: 'Simulation for economic evaluation of advanced manufacturing'. Proceedings of 1st International Conference on Simulation in Manufacturing
2. BUCHEL, A.: 'Comparison of design methodologies, characteristics and deficiencies'. Production Management Systems, IFIP, 1984
3. COWAN, D. A.: 'FMS for next generation PCB assembly'. Robot Assembly of PCBs, IProd.E. Feb., 1987

4. DU FEU, P. H.: 'Purchasing CAD as part of an integrated design and manufacturing system'. Proceedings of Internepcon, 1981
5. JOHNS, A. E.: 'Why simulate'. I.Prod.E. Seminar on Simulation, Feb. 1986
6. KINGSTON, M. R.: 'Road mapping the integrated CAE environment'
7. LOONEY, M. W.: 'Human factors in the introduction of CADCAM'. Design Conference, 1984
8. O'GUIN, M. C.: 'Constructing a strategic operations plan'. CIM Technology, Fall 1986
9. WARBY, A. H.: 'Choosing and using a simulation system'. Automan, 1985
10. YEOMAN, P. H.: 'Improving quality and productivity in systems modelling using the IDEF methodologies'. Micromatch, 1985
11. YEOMAN, P. H.: 'The IDEF Methodology'. Micromatch, 1985
12. 'ICAM Program Prospectus'. US Air Force Materials Laboratory, 1979

Chapter 10

Typical installations

Case-study material is invaluable in providing ideas and comparative data. This chapter gives a comprehensive summary of two companies, examining some of their basic attitudes, strategies and methods.

10.1 Introduction

In the preceding chapters we have considered the various constituent parts of methods of implementing advanced manufacturing in electronics. Examples of those companies which have successfully implemented AMIE are many. The two following case studies illustrate radically different approaches. They are both taken from reasonably large companies. In both instances the companies have manufacturing plants which often produce relatively small volumes of product of a wide variety. The lessons that can be drawn from these, in terms of approach, difficulties, required resources and business benefit, have wide application.

The first case study illustrates a company committed to computer integration. In its broadest sense this does not imply automation; however, this particular company has also installed a considerable amount of automation in direct assembly and materials handling.

The second company has not initially attempted to automate. Its approach was more concerned with establishing work flow and resolving organisational problems. In particular, the technique that they often applied was to establish how to automate the production of an item, identifying and eliminating problems. This had a very substantial effect in removing many of the original requirements for automation and in providing product that could be reliably assembled manually. With both companies, whether the approach was to integrate, automate or organise, one of the overriding considerations was that of ensuring quality throughout. Indeed, in one instance a defect level of 5 parts in 10^6 in the printed-circuit-board assembly area was easily achieved. Such achievements have a radical impact upon the organisation of work, the capability for flow and the level of re-work.

The different approaches taken by the two companies can be summed up very easily:

Company 1: Contain and control the problem by computerisation and automation.
Company 2: Identify and deal with the cause of the problem and the problem disappears.

10.2 Case study 1

10.2.1 Background

This large multi-national company has approximately one-third of its employees directly concerned with manufacturing. The number of people involved in manufacture has decreased by about 30% over the past three to four years, while sales volumes have increased and inventory levels have decreased. The company has very successfully introduced new products which are cheaper to build, more reliable and which offer a radically improved specification. By use of CIM techniques they have reduced inventory levels by about 30%.

The choice of the location of manufacturing sites is not limited by purely commercial factors. The company has intentionally considered social implications, and has sited some of the manufacturing plants in areas which could be considered as far from ideal. Despite the image of high-technology manufacturing plants being bright airy modern buildings, in this case study one of the three plants utilised an existing set of five old brick buildings. These buildings, with little scope for alteration or extension, do not appear to be ideal for a modern manufacturing operation, with the restrictions of height, closely spaced support pillars and narrow door ways. Despite these restrictions the company has very successfully implemented a CIM facility.

Three sites of this particular company are discussed, each with between 600 and 800 people. The first site is primarily concerned with design—design for manufacture and the development of manufacturing processes—while the other two are manufacturing plants.

10.2.2 Design and process development group
Outwardly a very high level of integration is evident. Every engineer within this group has access to a computer terminal linked into a totally integrated network. In addition to the computer integration, there are the underlying attitudes and thinking patterns which have greater significance. Employees are strongly aware of the implications of manufacturing changes which are to be expected over the next five years. A main thrust has been identified to reduce overall product cycle times and move towards batch sizes of one. In a computer-integrated environment there are obvious implications on inventory control if these objectives are to be achieved.

Within the group there are three main points recognised as having overriding significance:

- The company cannot be successful in the application of technology unless process development work is done prior to product introduction.
- The technologies of design, manufacturing, test and management tools must be integrated before they can be effectively applied.
- Technology developments result in positive value to the company only where there is a positive impact in a product design or manufacturing operation.

Having some basic goals and judgment criteria such as these obviously makes it easier to keep the company on course. One of the key computer tools used is simulation, carried out by both design and process development engineers.

Design simulation is used to help ensure total quality, and hence products that can be readily and reliably introduced into manufacture. Design simulation is not restricted to traditional logic and analogue simulation and timing verification. Use is also made of thermal simulation. Furthermore, PCB layout tools are integrated with mechanical packages to ensure correct packaging and the elimination of traditional mechanical tolerancing problems. The system provides direct links into the PCB manufacturing and assembly facilities. These include direct laser-imaging systems for the production of PCB prototypes, which may be up to 22 layers. This eliminates the use of photo-plotted artwork with all the associated problems of quality, physical document management and lengthened time scales.

Manufacturing process simulation is also carried out. The company recognises that simulation and modelling of existing and new manufacturing plants are essential if costly errors are to be avoided. These tools are used not only in connection with the introduction of new products, but also to resolve specific problems identified with existing products.

This group is also concerned with the cost justification of new technology. They recognise the need for major change in traditional accounting practices when considering the implications of modern manufacturing. The limitations of traditional accounting practices were identified, including the fact that in many instances traditional methods of justification would severely restrict the application of essential and beneficial new technology. They have therefore developed some very strong guidelines and rules for justification.

10.2.3 Manufacturing plants
The two manufacturing plants cited by this company both manufacture computer-based products. The fully computer-integrated manufacturing plants make extensive use of conveyoring, tote bins and automatic warehousing. All products are bar coded, and this is used as a data-entry method for the computer system in order to identify production routes, methods, programs etc.

Production is achieved by a combination of manual and automatic assembly, including the use of robotics where appropriate. In the case of manual assembly, all assembly information is presented to the operators via low-cost graphical computer terminals. These computer terminals, which are freely available to all manufacturing personnel, are used not only to supply assembly instructions, but can also provide reports concerning progress, quality, reliability and MTBF.

A considerable amount of effort has gone into the implementation of a closed-loop

MRPII system. This has recently gained class A status, which it has taken the company about 18 months to obtain. The system controls the following areas:

- Business planning
- Sales forecasting
- Production planning
- Master production scheduling
- Material requirement planning
- Capacity planning.

They have realised that obtaining class A status is not an end in itself. In order to maintain this level of excellence it requires continual commitment to develop, enhance and adapt.

Implementation costs for the MRPII system have been considerable. However, the apportionment of these costs is noteworthy. Hardware has accounted for just over 25% of overall system cost, with most of the remainder of the costs being accounted for by people aspects such as education and training.

The company has attempted to quantify the benefits arising from implementation of the MRPII system. This has been difficult to achieve because, concurrent with the MRPII implementation, has been company re-organisation, general education, training and the unrelenting drive to improve quality. However, justification for MRPII was based upon the following:

- Reduction in inventory   25—50%
- Reduction in overtime   50%
- Reduction in raw materials   2—5%
- Direct labour improvement   5—10%
- Indirect labour improvement   25%
- Improvement in customer service 10—20%
- Improvement in quality   10—20%
- Miscellaneous savings   10%

The benefits which have been achieved have in many cases far exceeded those estimated. For example, in practice, a seven-week product turn-around has now been reduced to six days. Many of the other benefits which have arisen come not only from the implementation of MRPII and associated automation tools, but also from a fresh approach to the design of new product. Design is now carried out with the aim of using less labour per unit. Dramatic savings in assembly time are now being achieved. For example, a unit which previously took 16 hours to complete is now taking only 16 minutes.

Although in many companies automated materials handling and stores, especially for work in progress, are seen as restricting factors, this company has achieved successful implementation of these facilities. Stores are very effectively utilised because of the high degree of planning and control exercised within the plant. For example, the average wait for each unit in the work-in-progress stores is a mere 2½ minutes.

A final key factor associated with the drive for quality was the development of strong links with component suppliers. There is continual positive feedback on MTBF data, coupled with information on batch numbers, dates etc. These relationships are seen as being vital in order to increase unit MTBFs. The company also provides graphical analysis of faults to the suppliers, and on some occasions has also been actively involved in providing equipment for on-line transfer of this information.

10.2.4 Summary

The company has learnt much from its experience of implementing computer-integrated manufacturing. The overriding message is the importance of people to any implementation. Everywhere in this company people were aware of the implications, goals and objectives of the new technologies being implemented. It was recognised that the success of installation depended on the involvement of the people responsible for the facilities from top management level down to the operators on the shop floor.

Other important lessons which have been learnt have been in estimating resources. In the early days the level of management understanding was under-estimated. The time required to educate and train people in the new technologies, and to make them aware of not only the implications but the possibilities of new technology were also under-estimated. The need to have core people involved from concept to integration was recognised as being important, and the problems caused by staff changes through the period of implementation were minimised wherever possible. Learning has to be re-cycled through each staff change.

The estimation of the computer power required in order to implement CIM was difficult to make. In particular, it was apparent that the estimates were inadequate as utilisation of the plant went up. Finally, despite the use of planning tools, it was difficult to achieve right-first-time objectives and plans. It was found that, as objectives were achieved, the expectations and goals got higher, thus providing further motivation for advancement.

10.3 Case study 2

10.3.1 Background
This large multi-national company has a number of manufacturing sites worldwide. Manufacturing cost, including material, accounts for virtually 50% of sales cost, with R & D accounting for a further 10%. It has a manufacturing research centre which is funded by a levy of 1·35% on annual turnover.

This company is very different from the first one, because, although it firmly believes in the value of automation, it also places strong emphasis on organisational issues and resolving basic problems before trying to automate. Elimination of direct labour is not a priority, as labour accounts for between 1 and 5% of product cost. There is very strong emphasis on quality. Design for manufacture is also given a place of considerable importance within the company.

As with the first case study, this also reviews several sites. One is the manufacturing research centre and the others are manufacturing plants.

10.3.2 Manufacturing research centre
The company has several major research centres. This first site is concerned solely with manufacturing research. It has two main objectives:

- To provide an area of focus for interconnection methods and for factory automation systems
- To provide the required technology in order to give the company a competitive edge.

One of the ways in which the research centre operates is to set up a manufacturing system to resolve a specific problem and then to move the equipment out into an operating-division manufacturing plant. Invariably any such development work is done in conjunction with engineers from the manufacturing plant. Development of these new facilities is often very expensive. Much of the funding for the development comes from the levy rather than requiring the recipient company totally to fund the development.

Staff from the centre had carried out a considerable amount of comparative research concerning manufacturing developments within other companies. For example, they identified that using JIT could result in higher component costs, not uncommonly 20%, and that the production lines might only be utilised for 75% of the time. They recongised that a traditional approach to justification could prove to be a serious limiting factor when considering any new approach. Their justification attempted to quantify many of the indirect types of benefit discussed in Chapter 2.

The company was keen to use industry standards where applicable. However, they often found a need to create their own, because no industry standards existed or because the standards that did exist were not applicable to their needs. In particular, the use of MAP was not considered to be suitable because of the large amount of indigenous equipment used within the company.

The company was carrying out considerable developments in the area of surface-mount devices. They expect over a period of ten years to 1995 that the usage of surface-mount will grow 40-fold. In 1986 surface-mount PCBs accounted for nearly 40% of all production. A single machine type made in the UK, corporately specified, was being used for all surface-mount placement.

The company is also active with tape automatic bonding (TAB). Currently devices have 170 connections, but they are aiming towards devices with 1000. Printed-circuit boards currently have a 12 'thou' pitch for tracks. It is the aim to get this down to 6 'thou' within a very short space of time.

There are a number of continuing developments concerned with the assembly of printed circuit boards. At a basic level one of these is concerned with using computer graphics in a similar way to that used by the first company cited. However, the company is also active in the development of robotic work cells. They use standard robots but have developed in-house vision and parts-feeding equipment specific to their application. Currently they recognise that robotic systems can be very expensive: up to £250 000 per work station. Justification of this level of expenditure is difficult. However, they view it very much as an investment for the future because they see that manual insertion costs will continue to rise, whereas robotic insertion will continue to decrease in cost and will start to level off by 1995.

Automatic machine vision is seen as an important and integral part of advanced manufacturing facilities. A major application is in the robotic assembly cells where vision is used to ensure that component legs are correctly positioned, to identify orientation problems and ensure correct positioning of components. In the area of surface-mount assembly, vision is used to measure the thickness of solder paste. In this application a laser beam is projected at an angle to the board and its deflection is measured. Automatic machine vision has many other applications within the company. For example, it is used to analyse the printing and dot size of the matrix printers manufactured by the company.

10.3.3 Small-batch manufacturing plant
This site was typical of many manufacturing plants found within the UK. Typically boards were produced in small batches, anything from 50 to 100 off. The division had a wide variety of different types of printed circuit boards. The 1300 different assemblies utilised over 4000 different components. An analysis of the production showed that the normal 80/20 rule applied. About 80% of their total volume was accounted for by 230 different assemblies. This product range used about half of the overall range of components. The company set up a flow line containing the range of components, holding each of these in two Kanban bins. Each board being assembled was held in a carrier and moved down the line to be assembled. Assembly documentation specific to the board was attached to the carrier. The line, 160 ft long, was manned by ten operators, each of whom was responsible for a few feet of the line.

This manual system had achieved one of the prime objectives of a flexible manufacturing system: the ability to manufacture batches of one cost effectively. This flow line process offered many advantages over previous methods:

- Quality control inspection staff had been reduced from 15 to two people.
- All components for a Kanban bin could be pre-formed at one time.
- Set up time had been reduced by 90%.
- Production cycle time reduced from three weeks to two days.
- Work in progress reduced by 50%.

A more conventional approach to assembly was taken for products outside the scope of the flow line. This utilised a more conventional kitting approach and individual operators. This method was usually limited to old designs, or where the design had not been very well thought out and there were assembly problems. In addition, the company resolved some of the worst of its manufacturing problems, which were caused by poor or old designs, by contracting out the work.

10.3.4 Volume manufacturing plants
The key to the success of this company's manufacturing lies in two main factors: the degree of design for manufacture and the emphasis on identifying and removing the root causes of problems. A major goal was to reduce the amount of work in progress and buffer stock throughout the organisation, where the objective is to achieve a three-day delay between demand and shipment. Currently it is taking seven days to manufacture a 28-board product which contains over 12 000 different components. This time is split up as one day to assemble, one day to test individual boards and the remaining five to commission and soak-test the complete assembly.

Design for manufacture is given a leading place in the company. It is recognised that over 50% of the cost, quality and cycle time of a product is fixed at the design stage. Thus it is vital that this process is carried out correctly. In one plant the engineers were rotated every six to twelve months between manufacture and design, in order to improve the overall manufacturability of products. Design for manufacture is carried out in an informal fashion with close co-operation between people at all stages. Some of the main emphases are to:

- Minimise the part count
- Minimise the set-up time
- Orient all components in only one direction
- Make it obvious when parts are missing or incorrectly inserted
- Make dis-assembly easy
- Design for self alignment
- Reduce the number of different component types.

Design reviews are regularly held where material lists, board layouts and assembly methods are assessed. At present, computer modelling is not used for this.

An integral part of design for manufacture is design for test. The company found this very difficult, as it is a complicated process to assess the cost impact of various approaches.

The company-wide attitude towards automation was very evident within the manufacturing plant. Thus they recognised that a high level of automation was not always necessary. They organised operations effectively, identifying problem areas, and dealt with them before attempting to neutralise the effect of the problem by automation. This approach often offered a better and lower-cost solution.

Production utilised a pull approach, with Kanban techniques being applied. Wherever possible, stores were not used but components were held near to where they would be used. For example, adjacent to the integrated-circuit insertion machines there was a rack full of tubes of ICs. About six days' worth of stock was kept on the shop floor. The amount was predicted either from historical information or from marketing information. Kitting was not carried out.

Insertion equipment was designed with flexibility in mind. For example, with the axial insertion machines there were on-line sequencers, eliminating the need for kitting in the form of sequenced bandoliers. The IC insertion machine was equipped with 60 feeders for the most commonly used components. In addition, the equipment had been modified by the addition of a parts-feeding arrangement which would allow up to 600 different types of ICs to be fed. All PCBs were held in carriers of the same size. This, coupled with the wide range of components that could be held on line, allowed any type of board to be assembled automatically.

Odd-form components were inserted using semi-automatic assembly tables. Each assembly table catered for 90 different component types, and, in addition, there was an adjacent on line store with 240 different parts.

The company is moving towards bar-coding assemblies with their type number, and using this as a means of identifying the assembly to the particular processing equipment. For example, the flow solder process parameters are now automatically adjusted to the particular board being soldered by this method.

The company had embarked upon a concentrated attack on manufacturing inefficiencies. In three years there were spectacular results:

- Scrap reduced by a factor of 14
- Inventory reduced by a factor of 21
- Floor area halved
- Work force halved.

This company had invested in an automated warehouse and conveyor system some time ago. However, they were finding that this could not cope with the number of transactions that were now required in order to support manufacture. They are planning to scrap the system as soon as possible and move to a local storage system adjacent to the assembly equipment. This policy was in keeping with their overall emphasis of simplification of the process, and the identification and elimination of basic problems. They pointed out that complex material tracking systems are not necessary if the production cycle time is short and problems are dealt with as soon as they are identified.

Thus the company was very successful in its ability to bring complex new products out into the market with a development time cycle of less than one year. The great emphasis given to design for manufacture and control of the process resulted in very high quality of production. Typically on the printed circuit board lines defect levels were at a level of 5 parts in 10^6.

10.4 Summary

These two case studies serve to illustrate two radically different approaches to modern manufacturing. Both companies are profitable in operation. They are able to meet the demands of the customer on time. They are able to introduce new products quickly. Their production quality is high. However, the approaches which they take are radically different. Perhaps the two approaches are best summed up in the sort of catch phrases commonly heard at conferences: *Automate or liquidate* and *Eradicate or liquidate*

These two case studies are, however, a snapshot at a particular time, and there are underlying trends in both which seem to be bringing them to a single converging point. For example, although the second company places much emphasis on identification and eradication of problems, it also applies automation where beneficial. This is seen, for example, in the manual flexible assembly line where currently assembly instructions are held as paper documentation. However, the company is working on graphical methods for displaying this information so that the inefficient paper work could be eliminated.

While case-study material is always interesting, it is of little value unless the lessons are applied. The fact that these examples are drawn from large companies in no way precludes the application of some of the methods and approaches to quite small companies. There is no reason why a company cannot place great emphasis on design for manufacture. At a very basic level, why not, for example, orientate all polarised components the same way on a printed-circuit board? Assembly tasks are simplified, quality is increased and complete visual inspection of all components on a board for orientation can be carried out in seconds rather than, perhaps, many minutes. Attention to basic details can have a profound effect on the operation of the company. Designers must recognise that their task is intimately related to manufacture, and that cost-effective production can only be engineered in at the design stage.

These two case studies served to highlight some important lessons. It is evident that many companies are investing large amounts of money to try to resolve the effects of problems rather than to identify and eliminate their root causes. In order to face up to competition and increase in profitability, bold and radical action is needed. Basic problems must be resolved; operating practices and traditional attitudes which hinder progress and profitability must be challenged and dealt with. Thus, as these two case studies illustrate, released from internal and debilitating pressures and problems, a company is able to respond to the market effectively.

Chapter 11
Component technology

Pertinent to AMIE is component technology. This is a rapidly developing area. Two significant developments are seen in the areas of surface-mount technology and custom silicon. These technologies are described and the consequences discussed.

11.1 Introduction

The definition of a manufacturing strategy is a complex task. It is not only concerned with how to manufacture. Inextricably linked to methods are the materials and components to be used. It is of little value applying advanced techniques to manufacture product, where outdated components make the product unattractive to the market place.

The rate of component development is exponential in nature—it is only 30 years since the transistor became a commercial reality, and 15 years since basic integrated circuits began to gain widespread acceptance. Today, the available technology enables the use of custom VLSI circuits by even small-scale users. Such developments have had tremendous impact on products and companies alike.

Today, two developments are particularly pertinent to the electronics industry: surface-mount technology (SMT) and application-specific integrated circuits (ASICs). This chapter will review these two substantially different technologies—the first relating to how electronic components are packaged and the second as to the nature of the components themselves. Properly applied, both these complementary technologies will provide substantial competitive advantages:

Decreased cost of raw material and manufacturing through simpler processes, fewer components and reduced product size
Decreased time to market: The only way to develop products using these technologies is through the use of CAE, which can provide a substantial reduction to the development time scale
Increased realiability through radically fewer discrete electrical interconnections and higher resistance to mechanical damage and electrical stress
Enhanced product specification: more functions, smaller size, lower power consumption
Simpler production process as a result of fewer components and assembly stages.
Improved electrical performance through shorter connections and greater parametric control of designs.

11.2 Surface mount

11.2.1 Surface mount: The technology
Surface-mount technology (SMT) is an approach to the packaging of electronic components whereby they are attached directly to the metallisation of the substrate, usually a printed circuit board.[7] Some of the more common styles of components are illustrated in Fig. 11.1. Surface-mount technology is gaining widespread acceptance, and rapidly making substantial inroads into the domain of the traditional leaded component and through-hole PCB technology.

The ramifications of SMT are considerable because the production process cannot be considered as an isolated entity. A coherent strategy is required for SMT. This will span design, procurement, assembly, test and rework. Assurance of quality, rather than testing out of defects, must assume a significant role. In particular, there will be a need for increased synergy between design and production, with automated tools for design and assembly growing in applicability, effectiveness and importance.

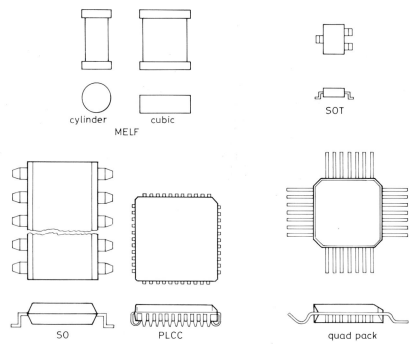

Fig. 11.1 *Some typical SMT package outlines*

11.2.2 Surface mount: The benefits

The benefits of surface mount primarily arise from three main factors:

- Miniaturisation
- Reliability
- Rationalisation.

These all relate to the components themselves, the substrate upon which they are mounted and the production techniques by which the devices are affixed to the substrate. Benefits include reduced costs and improved specifications:

- Surface-mounted devices (SMDs) are smaller than leaded-component equivalents. This results in smaller PCB sizes and a higher packing density, with the final product either being smaller and lighter or providing more functions per unit volume.
- Elimination of leads reduces the number of joints, and therefore helps increase reliability.
- Lead cutting and bending during assembly is not necessary, thus eliminating a source of stress and damage.
- Smaller devices and shorter leads results in reduced parasitic capacitance and increased operational speed.
- Low capacitance values can be achieved to a closer tolerance.
- SMDs offer scope for a considerably higher number of pins than with conventional packaging techniques.
- Devices, being small and with less material and constituent parts, cost less to manufacture.
- Substrates can be smaller and have fewer holes, thus reducing cost and increasing reliability.
- Substrates can be backed with thermally conductive material to readily and cheaply provide heat sinking for circuits, dissipating large amounts of power.
- Assembly equipment is smaller, faster and cheaper. Usually many more components can now be held on-line, resulting in reduced set-up times and greater production flexibility. One machine can now be used to place the range of components that would have previously required two or three automatic machines and some manual assembly effort to insert. This results in decreased capital cost, lower WIP levels and simpler production planning and control.
- Component stores can be very much smaller and located locally.
- A synergistic approach to the technology can result in very high production yields, making dramatic savings in the quantity of re-work.

11.2.3 Surface mount: Market trends

Market surveys show that surface-mount technology is replacing leaded-component packaging technology for existing product types. It is also creating new product and market opportunities across the complete production spectrum. Supplier information and independent surveys indicate a demand for SMDs which is tripling every three years, and which will account for 50% of all applicable component types by the early 1990s.

11.2.4 Surface mount: Impact on design

Whilst the need to design for manufacture has always been present, it becomes even more important with surface mount. Surface mount impacts upon design in a number of ways, including the selection of components, assembly and soldering methods, and test and re-work. Successful implementation of surface-mount technology can occur only when the manufacturing and test requirements are catered for during design.

Component selection: Automatic assembly is one key element in successful surface-mount implementation. Rationalisation of component usage can increase production flexibility. Attempts at component rationalisation have often been hindered by a legacy of old designs. A radically different set of components allows rationalisation without this concern. Specification of components is more critical because electrically identical components may vary physically in size, style and metallisation plating type. Standards must be set and adhered to in order to facilitate production flow.

Design methods: design rules concerned with pad shapes, sizes and connection methods are important and are related to soldering method. Manual design methods have always been open to error and abuse, with design rules being difficult, if not impossible, to implement. Increasingly CAD is seen as the only satisfactory way of ensuring that the manufacturing requirements, or design rules, are adhered to.

11.2.5 Surface mount: Substrate requirements

Surface mount brings with it more exacting requirements for substrate specification, because of the effect of mismatched thermal expansion coefficients between component and substrate. With leaded components any differential in these coefficients would usually be taken up by the component leads themselves. This is not the case with leadless components.

As a general rule, any ceramic leadless SMDs, with an edge longer than 6 mm, should not be used on paper- or glass-fibre based PCBs. Devices longer than this should only be used on a substrate where there is negligible differential in expansion coefficients such as phenolic glass, steel, aluminium and polyimide. Leaded SMDs, such as SO or quad packs, generally have sufficient lead flexibility to cater for any thermal mismatch.

11.2.6 Surface mount: Component sourcing

The lack of standards in the field of SMDs presents real problems for many users. Depending on the supplier, electrically identical devices may be physically different. In addition to the component itself, the method of supply and exact specification of, for example, the supply tape will be subject to variation. Factors such as these will, in certain instances, preclude any form of second sourcing for components. This is not, however, at variance with the concepts of JIT, and should not necessarily be viewed as a disadvantage of surface mount.

11.2.7 Surface mount: Production

This is reviewed in some depth in Chapter 4. Two factors are, however, pertinent at this point. The small size of components and lack of marking makes manual assembly unattractive, or even inapplicable, because of problems of mechanical alignment and identification of devices once unpacked. PCB cleaning can be difficult once large SMDs are in place; ultrasonic methods may be employed where other factors do not preclude it. Solder joints can be difficult to inspect, and therefore process quality is of paramount importance if yield and reliability are to be ensured at minimum cost.

11.2.8 Surface mount: Test strategy

Surface mount required a fresh approach to the subject of quality assurance.[14] The traditional approach for leaded component assemblies has been to rely on in-circuit test systems to identify faulty components and manufacturing defects. Following on from this basic level of inspection testing comes product functional testing, either using automatic test equipment

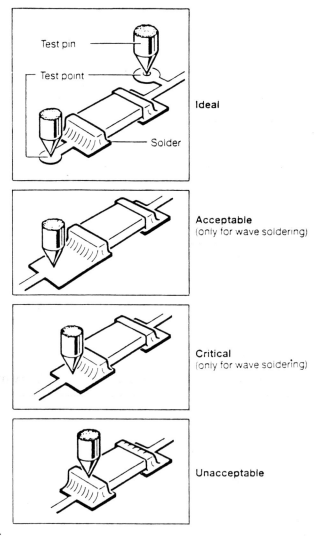

Fig. 11.2 *Test-point positions*

or a product system test approach. Whereas with leaded components a bed-of-nails fixture provides a relatively simple and cheap method of test-system interfacing, the reduced physical space associated with surface-mount devices and the lack of leads now makes this method complex and expensive, as Fig. 11.2 illustrates.

Strategy should centre on how to eliminate in-circuit test, aiming to ensure high yield by:

- High quality level of incoming components.
- Regular equipment maintenance.
- Enforcement of PCB layout rules.
- Minimum inter-process storage and handling.
- In-process inspection.
- Process control.
- Education and training.

11.3 Application-specific integrated circuits

11.3.1 ASICs: Introduction

Application-specific integrated circuits (ASICs), i.e. integrated circuits designed and manufactured for a specific application, are now making significant inroads into the domain of general purpose integrated circuits. The increasing power of low-cost computer-based design tools, complemented by low-cost production processes for prototypes, has brought the technology out of the specialist field into a position where it is attractive for a wide variety of companies, markets and products. ASICs offer a very substantial competitive advantage over designs based on standard ICs. Furthermore ASICs allow the creation of radically new products which would not otherwise be possible.

Generally the ASIC family divides into two

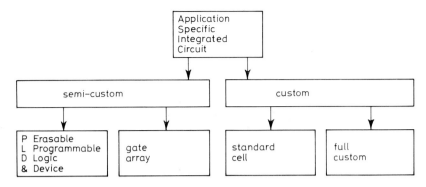

Fig. 11.3 *Custom silicon—the options*

Table 11.1 *ASIC parameter comparisons*

Device parameter	PLD/EPLD	Gate array	Standard cell	Full Custom
Design cost	Very low	Low	Medium	High
Design cycle time	Hours—Days	Days—Weeks	Weeks—Months	Months
Complexity-gates	Very low <2000	Low <20 000	Medium—High	Medium—High
Volume unit costs	Very high	High	Low	Very low
Design security	Low	High	Very high	Very very high

sub-sections—semi-custom and custom. Splitting these down further, as shown in Fig. 11.3, are programmable logic devices (PLDs), gate arrays, standard cell and full custom.

There are wide variations between the devices, covering technology limits, design time and cost, and manufacturing cost. Product and market variations will affect the choice of device type. Figs. 11.4 and Table 11.1 summarise some of the relevant factors affecting choice. In addition, circuits can be fabricated in a number of ways, using, for example, bipolar or CMOS technology. The choice is usually application dependent, with the speed/power benefit of bipolar being offset by increased size and dissipation.

Use of ASICs can provide some very substantial advantages over the use of standard ICs:

Performance, in terms of speed and power dissipation, is improved through reduced size and increased integration.
Reliability is improved through radically fewer solder joints, fewer components and simpler assemblies.
Design security is maintained because a design implemented in silicon cannot readily be copied by a competitor.
Design costs and cycle times can be very short. For example, it is quite feasible to be able to design a circuit, simulate and verify its operation and lay out a 300-gate IC in less than a week. Prototypes can be available in a further week.
Prototype costs can be low.

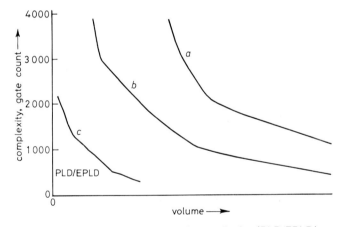

Fig. 11.4 *Type by volume and complexity (PLD/EPLD)*
 a Full custom
 b Standard cell
 c Gate array

Balanced against the advantages are potential problems induced by poor design. Unlike a PCB, an IC once manufactured cannot be modified. De-bugging is limited to the external access points—there is no scope to probe internal circuit elements, cut tracks or add wire links as there is with a PCB. Design should only be carried out using CAD. Well

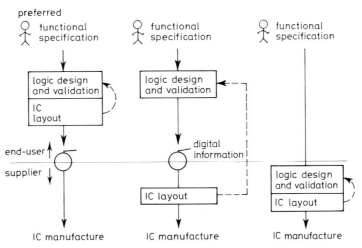

Fig. 11.5 *ASIC design options*

used, a CAD system should enable right-first-time designs to be achieved as the rule rather than the exception. CAD was discussed in more detail in Chapter 3, though the following is a brief over-view of the major steps of the design process relating to ICs.

Circuit entry tools will ensure the electrical validity of the circuit and simulation tools enable the functionality and timing of the circuit to be evaluated and validated. The use of such simulation tools is essential in order to achieve good designs first time.

A vital part of the design cycle is ensuring that testability requirements are met. Simulation tools can be used to verify that the design can be tested and also that the test program is efficient. Production IC testers costing many millions of pounds have to be used effectively, and this often cannot be achieved unless testability has been built in to the design. Building in the test requirement may well require additional circuitry, which at the initial design stage can usually be added quickly and cheaply. Layout tools are then used to create the physical design, using circuit information already captured.

Parameters such as timing will change from the theoretical value once a device is laid out. This is due to the effect of track capacitance. Timing verification must therefore be carried out again, after layout is complete.

Design activities can be carried out in a number of ways, though all utilise CAD. Fig. 11.5 illustrates possible options. The preferred method is where the end user has total control over the design. Powerful low-cost systems, currently costing less than half a year's salary, make this a very attractive option even for users with minimal design requirements. The layout stage is not a draughting function; neither is it necessarily one that should be carried out in isolation by a specialist IC designer. The sort of tool described should enable a circuit designer to readily acquire the skills of IC design, bringing with it the benefit arising from interaction between the two closely related phases of design.

Once designed, devices have to be manufactured. Data is transferred from the CAD system to the manufacturer in digital format. This data may already have been post-processed to create the machine-drive files required for mask creation or E-beam lithography, or it may be in a format such as EDIF.[4] In the latter case the manufacturer would process the data, converting it to the required format to drive his equipment.

11.3.2 ASICs: Device descriptions

The method of design and manufacture varies considerably between the various devices, as shown in Table 11.2, where at one end of the spectrum the complete device is customised, to the other where the device, as supplied, is a standard part.

11.3.3 Programmable logic devices

Programmable logic devices (PLDs) and eraseable programmable logic devices (EPLDs) consist of programmable arrays of gates. In order to customise the device, the user has to program it in a very similar way to that used for programmable read-only memories (PROMs). The standard silicon is therefore customised, to provide the ASIC, by a post manufacture operation. A typical device, at the upper limit of currrent technology, has 68 pins, and provides the user with an

Table 11.2 *ASIC device design and fabrication*

Design and fabrication		ASIC type			
		Semi-custom		Custom	
		PLD/EPLD	Gate array	Standard cell	Full custom
Logic cell	Design	Standard	Standard	Standard	Custom
	Placement	Standard	Standard	Custom	Custom
Silicon-wafer fabrication	Active layer	Standard	Standard	Custom	Custom
	Interconnection layer	Standard	Custom	Custom	Custom

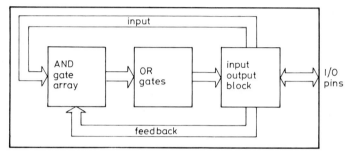

Fig. 11.6 *Typical PLD/EPLD block diagram*

equivalent gate count in excess of 2000. This device comprises a number of cells. Each is split into AND gates, OR gates and an input/output (I/O) block, as Fig. 11.6 shows. Programming of the AND gate array allows the functionality of the device to be configured to suit the application.

The significant advantages of PLD/EPLDs lie in the very low design cost and the very rapid route to finished product, measured in minutes as opposed to days. Against this is offset higher unit costs arising from programming costs, and the intrinsically higher basic component costs which result from a very poor silicon utilisation.

PLD/EPLDs find significant application either in low-volume or low-complexity applications and also where time to market is critical. They can also be used as a proving device prior to commitment to true custom silicon. As examples of what the devices can perform, a 1200-gate device could be used to implement a 3½-digit display driver, while a 700-gate device could achieve bus arbitration and control logic functions for a microprocessor system.

11.3.4 Gate array

Gate arrays, or uncommitted logic arrays (ULA), account for a major part of the custom silicon market size. It is estimated that, as shown in Fig. 11.7, some 75% of all custom silicon is accounted for by these devices. As gate arrays consist of a regular array of standard cells, they can include interface and analogue functions with the logic cells.

The silicon wafer is standard or uncommitted until a very late stage of manufacture, with the customisation being achieved by one or more final interconnection layers. Interconnections are made by means of an aluminium conductor layer, which is etched away in a very similar manner to the copper track on a PCB. During the customisation process, the photoresist covering the aluminium is exposed, either to the required pattern through a mask, or the pattern may be written directly onto the resist by means of electron-beam (E-beam) lithography. In the E-beam process, an electron beam is modulated and deflected in a very similar way to that used for TV-type cathode-ray tubes. E-beam allows a very rapid method of providing devices in low volume

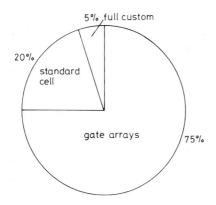

Fig. 11.7 *Custom silicon—type by market share*

because it obviates the need for mask production and proving.

Generally the volume IC manufacturers have little interest in very low-volume users, and may not provide adequate facilities for prototyping. However a number of smaller companies can offer these low-volume (<100 off) and prototype (1–10 off) production facilities, usually using silicon brokerage. Silicon brokerage allows several different designs to be manufactured on the same silicon wafer, rather than dedicating a complete wafer to a single design.

Gate arrays have widespread application. They are often used as prototypes that will eventually be developed as standard cell or full custom. Large designs may be prototyped as several smaller gate arrays which can subsequently be merged. This fast and economic route to custom ICs has enabled ASICs to be used in a wide range of products, suiting a cross-section of companies and markets. Enormous economic and performance benefits have given outstanding successes in a diverse range of products and markets:

- Telecommunications: handsets to exchanges
- Computers: mainframe to personal
- Toys, TV games
- Power tools, domestic appliances
- Photography
- Security systems, locks
- Industrial control
- Vending machines
- Medical
- Automotive
- Aerospace.

11.3.5 Standard cell and full custom
These two categories of component vary very considerably from gate arrays because the whole of the silicon die is now specific to the device. During manufacture eight or more custom masks are required.

Standard cell may be related, at a very simplistic level, to PCBs. With a PCB the designer has a library of components which can be selected, interconnected and laid out to give the required functionality. The standard-cell approach is based on having a library of standard building blocks for both circuit and physical layout. These can be selected, interconnected and laid out in a similar fashion to a PCB. A wide range of standard cells are available. These include standard logic functions such as 74-series gates, flip–flops, counters etc., and also super- or mega-cells such as microprocessor cores, which are complete LSI/VLSI devices in themselves, and the term 'macro-cell' is often used in the context of such functions.

'Full custom', as the name implies, allows full customisation, and requires the designer to design right down to the basic transistor level.

These two approaches obviously have many cost and time implications, as Table 11.1 shows. Balanced against these are the benefits arising from a higher degree of integration. The complexity of devices tends to be higher, with custom devices going up to gate counts in the hundreds of thousands. These devices are usually aimed at either very specialist applications, e.g. computer-processing elements, or at large volume markets. Volume unit costs are reduced because of the improved silicon utilisation which arises out of the virtual total control over not only the layout of the device, but also on the basic die size.

The distinction between standard cell and full custom is somewhat artificial because it is related to design methodology rather than having any physical significance. This distinction is becoming increasingly blurred as the power of CAD tools increases.[1,10] A variety of tools have been developed by different vendors and they are generally described as silicon compilers, or the term 'soft macros' may be used.

A silicon compiler is a parameterised cell generator and provides the opportunity automatically to create a cell specific to an application. In use, the designer selects the function to be compiled and specifies the appropriate values of parameters. From this information the compiler will create the required cell. The most common types of functions available are memory arrays, arithmetic elements and shift registers. For example, if a memory array is required, the designer would need to specify the memory-element type and the array size, leaving the compiler to generate the required circuit description and physical layout.

Such design tools allow individual blocks or cells to be designed, verified, laid out and stored for re-use as a proven circuit. During layout phases the complete cell, which may itself consist of thousands of gates, can be moved around as a single element and interconnected with other cells or output pins. Keeping functional blocks as discrete elements on the single complete die has tremendous benefit in allowing the design activity to be partitioned, thus easing the design task because it can now be carried out in smaller and better defined steps. Benefit also arises from the re-use of information.

Whereas gate arrays could be thought of as board-level replacements, custom silicon allows system-level replacement. This has significant impact on the approach taken to development. It will necessarily require a tighter initial functional specification and exhaustive design simulation and verification. Testability must be built in from the concept of the design.

11.4 Summary

The component technologies which have been described, when implemented in a well managed fashion, will provide:

- Cheaper products
- More reliable products
- Smaller products
- Better specified products
- Products out sooner
- More effective utilisation of key staff
- Increased utilisation of production area and equipment.

All these factors are vital to keeping companies, both large and small, well placed in the competitive world-wide race. Perhaps most important, the use of the appropriate technology opens new horizons, enabling products to be developed which would never have been conceivable in the days of standard devices and leaded components.

11.5 Bibliography

1. AMBLER, A. P. *in* Proceedings of the Third Silicon Design Conference (Electronic Design Automation Ltd.)
2. BEST, S.: 'Tackling SMT with a minimum of pain', *Electron. Manufacture and Test*, Feb., 1987
3. BOYCE, A. H.: 'ATPG for scan testable design (Marconi Research)
4. CARLSTEDT-DUKE, T.: 'Experiences using EDIF' (Daisy Systems UK)
5. CURTIS, W.: 'Silicon compilation—A revolution underway.'
6. HICKS, P. J. (Ed.): 'Semi-custom IC design and VLSI' (Peter Peregrinus, ISBN 0 8641 0111 1)
7. HINTRINGER, O.: 'An introduction to surface mounting' (Siemens Aktiengesellscahft)
8. HORSLEY, S.: 'Surface mount forum', *Electron. Production*
9. MAIWALD, W.: 'PCB layout recommendations' (Siemens Aktiengesellschaft)
10. MAVOR, J.: 'Introduction to MOS LSI design' (Addison–Wesley, ISBN 0-201-14402-6)
11. PERRY, C.: 'The impact of SMDs on CAD (Academi Systems)
12. ROWE, G. L.: 'Setting up a surface mount facility'. Proceedings of the SMT Conference, Long Beach, CA
13. 'Circuit board test systems' (Markt and Technik)
14. 'Design to testabiliy—A must with SMDs' (Markt and Technik)

Chapter 12

A look ahead

Technology is never static. This chapter looks at some factors which may colour the future, and therefore the way in which technology may develop.

12.1 Introduction

The technological revolution, unleased when man first tamed the power of fire, discovered how to make tools and developed the wheel, continues. It forges ahead at an exponential rate. The sum of the future developments in advanced manufacturing until the end of the century is expected to exceed the sum of those experiences to date. Computer-based tools and knowledge that already exist constantly fuel further development, creating an environment where survival becomes dependent on effective utilisation of technology. To stand still, or even delay unduly, can rapidly lead to a position where it may be too late to catch up. The future is certain to be one of ever-increasing change, prolific advances in technology, and constant challenge and excitement.

There are many driving forces which will shape the future, and Fig. 12.1 reflects some of these. They fall into three main groups: competitive factors, technology and people aspects. It is outside of the scope of this book to consider the latter in anything but a cursory manner. However, factors such as the ability of society to adapt to changing work patterns, political forces which drive defence-oriented development, growing problems of world debt and uneven global distribution of wealth and resources will all colour the future.

This chapter briefly considers competitive elements before considering some of the technology trends which are already evident.

12.2 Competitive pressure

The increasingly discerning market place is constantly demanding new and better products. It is those companies which have learnt to respond to such requirements that will be successful. Two of the requirements illustrated in Fig. 12.1 are highlighted: new products and increased value. The scope for new products is bounded only by man's imagination. Constantly companies are developing new market areas. Furthermore, the introduction of rad-

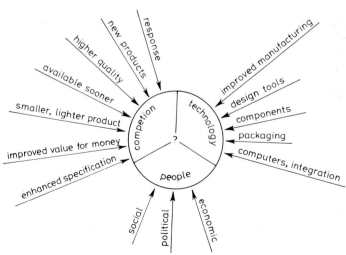

Fig. 12.1 *The future—driving forces*

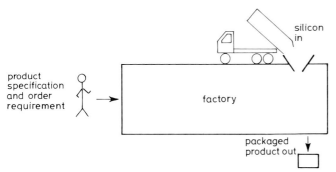

Fig. 12.2 *Factory of the future*

ically different and innovative products can create new markets and additional demand.

Minimum cost is not always the major differentiating factor for choice; increasingly the market is demanding improved value for money. This may be reflected in obtaining products with an improved specification, rather than reducing the cost of existing product. The application of new technology will often lead to an enhanced specification at lower cost. Competitive forces such as these will drive companies towards improved efficiency and responsiveness. These are achieved through increased levels of integration—both of equipment and functions—a greater utilisation of the available technology, and improved business methods.

12.3 Technology of the future

Industry is at the start of a very long introduction curve of CIM, which will stretch beyond the year AD 2000. The factory of the future depicted in Fig. 12.2 is still some years off.[4] However, trends are evident which will make it possible. Expert systems and artificial intelligence will convert customer requirements into manufacturing methods. Flexible automation will transform raw material into finished product, reliably, rapidly and cost effectively.

Already, in the mechanism field, the Alvey sponsored 'design to product' demonstrator is moving towards implementing a demonstration system similar to that illustrated in Fig. 12.2. Changes will be farreaching, and will span the entire business.

12.3.1 Design
Computer-aided design and engineering is moving towards computerised design and engineering. No longer will the skill of the designer be required to direct the design process. Expert systems and artificial intelligence within these will automatically translate user requirements into manufacturing information. An obvious early example of this approach is in silicon compilation.[2,3] Automation of the low-level, repetitive and error-prone tasks, associated with silicon layout, allows the system designer to concentrate on higher levels of design and the more intricate areas. This allows a greater amount of design time to be used to test ideas at a physical level and to allow 'what if' questions to be asked. A tool offering similar features will be developed for PCBs. It will have the capability for automatic design of products which can be manufactured reliably and cost effectively.

Such tools will merge into a single work bench, allowing the system designer to enter this thoughts at a very high level. Designs will be automatically partitioned into the appropriate technology. Automatic tools will create 'right first time' designs, and produce all the manufacturing documentation.

12.3.2 Manufacture
The evolutionary developments under way in production equipment will continue, resulting in equipment which will assemble products more reliably. Cycle time may not necessarily undergo a dramatic decrease, though effective throughput will increase as program data is automatically downloaded from CAD, set-up tasks are eased and assembly quality improves.

A more revolutionary change will be seen in the application of robots to electronics assembly. The high degree of flexibility claimed for robots is rarely seen in existing installations. However, this is changing, as Fig. 1.2 shows. Such a system, linked into CAD, will automatically mass produce one-off products from over 2000 components. Developments in robot technology such as increased gripper flexibility, low-cost parts feeders, vision, greater computing power, improved communications, higher degrees of absolute accuracy, higher speed operation and a modular control architecture, will all contribute to the development of the long awaited flexible manufacturing cell.

12.3.3 Computers and integration
Computers, and the integration of computer-based systems, will form a significant part of tomorrow's advanced manufacturing. The power of computers, in terms of speed of operation and the method of programming,

will change, with a greater use of high-level languages, expert systems and artificial intelligence.

Larger programs require a new approach to system architecture: one approach is that based on functional decomposition. In this, a complicated problem is split into many thousands or even millions of fundamental functions. Automatically generated code, associated with each independent function, will be dynamically allocated to a free processor in a multi-processing computer system. Such an approach considerably eases the complex task of developing reliable software for the large systems required in a factory, such as the one illustrated in Fig. 12.2.

The task of integration is constantly being eased by the development and adoption of new standards. However, a single plug compatible standard is currently difficult to conceive because of the variety of application-specific requirements that will always exist.

12.3.4 Component and packaging technology
ASICs will rapidly become the accepted way to implement the majority of new designs. However ICs themselves are going to undergo a substantial revolution. Vertical structures and atom-thick silicon-on-insulator technology will break the 20×10^6 device-per-chip limit associated with traditional structures and fabrication methods. Further advances should yield geometries of the order of 300–500 Å and densities of about 100×10^6 devices per chip during the 1990s.

Device packaging is increasingly the subject of standards and tape automated bonding (TAB), is emerging as an important method. Companies are already talking of 1000-lead devices with leads on a 6 'thou' pitch. Needless to say, automatic assembly will be the only option for such a development.

PCBs will continue to provide an attractive method of mounting components at the system level; however, the pitch of tracks will continue to decrease; for example, PCBs with a track pitch of 1·6 'thou' are already becoming available. Techniques under development, such as hierarchical interconnection technology (HIT) cater for radical increases in packing density, whilst reducing unit costs and easing testability.[1]

The impact of such developments will be farreaching, affecting, for example, the design process. Bread-boarding will become inapplicable and any modification of the product impossible. A right-first-time approach to design, utilising simulation tools, will rapidly become the only possible option.

12.4 Summary

Many of the developments presented in this chapter may seem to be a huge leap away, and far removed from the reality of today's immediate problems. Closing the gap between the two will never be easy, neither will it be achieved with an 'instant technology' solution. Indeed, as the definition of AMIE presented at the start of the book illustrates, it is as much to do with the management of the technology as it is with technology itself. It is those companies who take hold of these developments in a well planned and managed manner which will prosper. They will grow and flourish, providing cost-effective reliable new and innovative products on demand and on time in an increasingly competitive market.

12.5 Bibliography

1. ANSTEY, M. J.: 'A three dimensional interconnection system', *Electron. Eng.*, March 1985
2. BENTLEY, M. J.: 'Chipcode', *Silicon Design*, July 1986
3. CURTIS, W. P.: 'Silicon compilation—A revolution underway'. Proceedings of the Third Silicon Design Conference, ISBN 9510 64402
4. DU FEU, P. H.: 'Purchasing CAD as part of an integrated design and manufacturing system'. Proceedings of Internepcon, 1981
5. LANCASTER, M.: 'Standardising surface mounting', *New Electronics*, March 1987
6. 'Application specific integrated circuits', *Electron. Weekly*, Nov. 1986

Chapter 13

Glossary of terms

ABCD classification: a classification method for ranking the operating effectiveness of manufacturing requirement planning systems.
AGV: Automatically Guided Vehicle.
AI: Artificial Intelligence.
Air knife: an air jet which is used on a flow-solder machine in order to reduce the incidence of solder defects.
AMIE: Advanced Manufacturing In Electronics.
AMT: Advanced Manufacturing Technology.
Analogue simulation: a computer tool for simulating the operation of an analogue circuit.
ASCII: American Standard Code for Information Interchange—a very widely used computer data format for textual information.
AS/RH: Automatic Storage, Retrieval and Handling.
ASIC: Application-Specific Integrated Circuit—a generic term covering PLDs, EPLDs, gate arrays, ULAs, custom and semi-custom integrated circuits.
ATE: Automatic Test Equipment.
ATPG: Automatic Test Programme Generation.
Auto-placement: a computer tool for automatically positioning components during the design phase.
Auto-router: a computer tool for converting a rat's nest into the routes for copper track on a printed circuit board.
AVI: Automatic Vision Inspection.
Back annotation: the task of reflecting back onto a circuit diagram any changes made to circuit notations during the PCB layout phase.
Batch mode: a method of controlling computer programs whereby instructions are taken from a computer file rather than interactively through a keyboard.
BOM: Bill Of Materials.

CAD: Computer Aided Design—usually schematic entry and layout.
CADMAT: Computer Aided Design Manufacture and Test.
CAE: Computer Aided Engineering—usually CAD and simulation.
CASE: Common Application Service Elements—a MAP protocol.
CIM: Computer Integrated Manufacture.
Clam shell: a bed-of-nails fixture used for simultaneous access to both sides of a printed-circuit board.
Code 3 of 9: a bar code in widespread use through the electronics industry.
Combinational tester: an automatic tester which combines the functions of in-circuit and functional testing.
CSMA/CD: Carrier Sense Multiple Access with Collision Detection.
Custom IC: an integrated circuit which is designed for a specific application.
Data dictionary/encyclopaedia: an index to a relational database.
Data-driven communications: communications which are controlled by the nature of the data being transferred.
DBMS: Data Base Management System.
DCF: Discounted Cash Flow—a financial-analysis methodology which accounts for the time value of money.
Decomposition: a method of splitting a problem down into small elements—usually used in association with modelling techniques.
DTI: Department of Trade & Industry.
E-beam lithography: a method of writing patterns using an electron beam. This is usually associated with creating the required pattern on the photo-resist for the final metallisation layer of a gate array.
EDIF: Electronic Design Interchange Format.
EPLD: Erasable Programmable Logic Device.
Event-driven simulator: a simulator that is

98   Glossary of terms

driven by the occurrence of an event rather than the passing of a pre-set period of time.

Fault simulation: a technique for simulating commonly found faults on printed-circuit boards in order to assess the scope of the test program and also to create a cross-referenced dictionary of faults and their associated effect.

FMS: Flexible Manufacturing System—a system which combines microelectronics and mechanical engineering in order to bring economies of scale to batch work.

Foot print: the area of a printed-circuit board required to cater for the insertion head or the lead cut and clinch mechanism of an auto-inserter.

FTAM: File Transfer Access and Management—a MAP protocol.

Gate allocation: the allocation of circuit elements to physical packages.

Glitch: a transient state change of logic data lines—usually caused by timing problems, race conditions or noise.

Hardware accelerator: another term for a simulation engine.

Hierarchical design: a technique where the design is carried out hierarchically—usually used in association with CAE systems where simulation may be carried out at a number of levels ranging from detailed circuit up to functional blocks.

HIT: Hierarchical Interconnection Technology—a component mounting and interconnection technology.

Hybrid circuit: an electronic circuit which is built on a substrate where conductors and insulators have been produced by conductive inks rather than normal copper track; resistive circuit elements are created through the use of an ink with appropriate resistivity.

IC: Integrated Circuit.

ICAM: Integrated Computer Aided Manufacturing programme—a programme initiated by the US Air Force.

ICT: In-Circuit Tester.

IDEF: ICAM Definition language—a set of modelling methodologies resulting from the ICAM Programme.

IEE: Institution of Electrical Engineers.

IGES: Initial Graphical Exchange Specification—a data-exchange format for CAD systems.

IR: Infra-Red.

IRR: Internal Rate of Return.

ISO: International Standards Organisation.

JIT: Just-In-Time—a production philosophy where a task is carried out just-in-time to satisfy the demand.

Kanban: a material-provisioning system used in conjunction with JIT implementations.

LAN: Local Area Network.

Library: another name for a portion of a data base which describes attributes of components. Usually used in conjunction with CAD, CAE and ATE.

LISP: a computer language.

Logic simulation: a computer-based tool for simulating the operation of a logic circuit.

LSI: Large Scale Integrated circuit.

Machine vision: a computer-based tool which is used to acquire an image, analyse it and produce definitive information according to the application; usually used in conjunction with inspection tasks or as a means of providing positioning information to automatic assembly and robotic systems.

MAP: Manufacturing Automation Protocol—a protocol for the interconnection of alien computer-based devices.

MELF: Metallised Electrode Face—an SMD where metallised faces provide the electrical connection, typically for resistors and capacitors.

Mixed-technology PCB: a printed-circuit board which has both surface-mounted and leaded components.

MMS: Manufacturing Message Service—a MAP protocol.

MRP: Material Requirement Planning.

MRPII: Manufacturing Resource Planning.

MTBF: Mean Time Between Failure.

Multi-layer PCB: a printed-circuit board which has internal conducting layers.

Net list: a list of electrical interconnections.

NPV: Net Present Value.

Odd-form component: a leaded component which cannot be inserted by the normal range of axial, radial and IC component insertion machines.

Off-line programming: the task of creating equipment programs away from the target equipment.

On-line programming: a method of creating programs for equipment whereby the program is generated on the equipment, thereby stopping production; this method is often referred to as teach programming.

OPT: Optimised Production Technique—a

planning technique aimed at identifying and eliminating bottlenecks.

OSI: Open Systems Interconnection.

Paperless repair: a computerised data-base system which automatically logs test faults, presents these to a re-work operator by use of a computer terminal, and captures re-work information on-line.

Parser: a computer tool which examines key words in data lists to assess whether the ensuing information has relevance to the task in hand.

Partitioning: the segmentation of a conceptual circuit into discrete physical entities.

PLCC: Plastic Leaded Chip Carrier—a form of SMD package.

PLD: Programmable Logic Device.

Power cycling: a method of carrying out environmental stress screening where the power is turned on and off.

PROM: Programmable Read-Only Memory.

Rat's nest: an illustration of the electrical connections on a printed-circuit board without any reference to the exact physical route.

Re-firing: a technique used for automatic assembly where, in the case of failure, a component is discarded and a fresh attempt made.

RIP: Raw and In Process material.

Risk analysis: an analysis of the risk involved in adopting a particular strategy or investment plan.

RS232: a digital communication standard primarily defining transmission rates and electrical signal levels.

Rubber banding: a graphical facility on CAD systems whereby, when a entity such as a circuit-diagram symbol or component on a printed-circuit board is moved, the graphical connectivity of the electrical signal is maintained.

Semi-custom IC: an application-specific integrated circuit which is designed using standard circuit building bricks.

Sensitivity analysis: a technique for evaluating the sensitivity of strategy and investment plans to changes in various input parameters.

Silicon brokerage: a manufacturing method for ICs, usually reserved for prototype and low-volume ASICs, whereby several IC types will share the same silicon wafer.

Silicon compiler: a computer design tool which will compile a circuit using parameterised circuit elements.

Simulation engine: a computer which is specifically designed with a particular aspects of simulation in mind.

SMC: Surface Mount Component.

SMD: Surface Mount Device, often interposed for SMC.

SO: Small Outline—an SMD package type.

SOT: Small Outline Transistor—an SMD package type.

SQC: Statistical Quality Control—a methodology for controlling quality through the analysis of performance statistics.

STEP: Standard for Exchange of Product Data—a common vehicle for transferring design data across an extremely wide range of engineering disciplines.

Stress screening: a technique for trying to eliminate early-life failures by subjecting a product to stress of various types.

Substrate: the base material upon which an electronic circuit is constructed.

TAB: Tape Automated Bonding—a method of component packaging.

Temperature cycling: a stress-screening technique whereby the temperature is cycled.

Thermal modelling: a computer technique for modelling temperature profiles of electronic circuits.

Timing verification: a computer tool for identifying any timing violations in a logic circuit.

TOP: Technical and Office Protocols.

Truth table: a table which defines the sequential relationships of output to input for a logic circuit.

ULA: Uncommitted Logic Array—often referred to as a gate array.

ULSI: Ultra Large Scale Integration.

UUT: Unit Under Test.

VLSI: Very Large Scale Integration.

Weibull curve: a graph, approximately shaped like a bath tub, which relates failure rate to passage of time.

What if?: an abbreviation for 'what would happen if this event occurred?'—usually used in conjunction with MRPII systems.

WIP: Work In Progress.

4GL: Fourth Generation Language.

80/20 Rule: a commonly found relationship between variables; e.g. 20% of all product types account for 80% of production volume.